Mehmet İsmet Can Dede

Fault-Tolerant Teleoperation Systems Design

Mehmet İsmet Can Dede

Fault-Tolerant Teleoperation Systems Design

Employment of Fault Tolerance Features and Virtual Rapid Robot Prototyping in Teleoperation Systems Design

VDM Verlag Dr. Müller

Impressum/Imprint (nur für Deutschland/ only for Germany)
Bibliografische Information der Deutschen Nationalbibliothek: Die Deutsche Nationalbibliothek
verzeichnet diese Publikation in der Deutschen Nationalbibliografie; detaillierte bibliografische
Daten sind im Internet über http://dnb.d-nb.de abrufbar.
Alle in diesem Buch genannten Marken und Produktnamen unterliegen warenzeichen-, marken-
oder patentrechtlichem Schutz bzw. sind Warenzeichen oder eingetragene Warenzeichen der
jeweiligen Inhaber. Die Wiedergabe von Marken, Produktnamen, Gebrauchsnamen,
Handelsnamen, Warenbezeichnungen u.s.w. in diesem Werk berechtigt auch ohne besondere
Kennzeichnung nicht zu der Annahme, dass solche Namen im Sinne der Warenzeichen- und
Markenschutzgesetzgebung als frei zu betrachten wären und daher von jedermann benutzt
werden dürften.

Coverbild: www.purestockx.com

Verlag: VDM Verlag Dr. Müller Aktiengesellschaft & Co. KG
Dudweiler Landstr. 125 a, 66123 Saarbrücken, Deutschland
Telefon +49 681 9100-698, Telefax +49 681 9100-988, Email: info@vdm-verlag.de
Zugl.: Miami, Florida International University, 2007

Herstellung in Deutschland:
Schaltungsdienst Lange o.H.G., Zehrensdorfer Str. 11, D-12277 Berlin
Books on Demand GmbH, Gutenbergring 53, D-22848 Norderstedt
Reha GmbH, Dudweiler Landstr. 99, D- 66123 Saarbrücken
ISBN: 978-3-8364-8896-9

Imprint (only for USA, GB)
Bibliographic information published by the Deutsche Nationalbibliothek: The Deutsche
Nationalbibliothek lists this publication in the Deutsche Nationalbibliografie; detailed
bibliographic data are available in the Internet at http://dnb.d-nb.de.
Any brand names and product names mentioned in this book are subject to trademark, brand or
patent protection and are trademarks or registered trademarks of their respective holders. The use
of brand names, product names, common names, trade names, product descriptions etc. even
without
a particular marking in this works is in no way to be construed to mean that such names may be
regarded as unrestricted in respect of trademark and brand protection legislation and could thus
be used by anyone.

Cover image: www.purestockx.com

Publisher:
VDM Verlag Dr. Müller Aktiengesellschaft & Co. KG
Dudweiler Landstr. 125 a, 66123 Saarbrücken, Germany
Phone +49 681 9100-698, Fax +49 681 9100-988, Email: info@vdm-verlag.de

Produced in USA and UK by:
Lightning Source Inc., 1246 Heil Quaker Blvd., La Vergne, TN 37086, USA
Lightning Source UK Ltd., Chapter House, Pitfield, Kiln Farm, Milton Keynes, MK11 3LW, GB
BookSurge, 7290 B. Investment Drive, North Charleston, SC 29418, USA
ISBN: 978-3-8364-8896-9

DEDICATION

I dedicate this book to my parents; Mustafa Ruhi and Nur Dede.

ACKNOWLEDGEMENTS

I am grateful for the support, good wishes and prayers of my parents, grandmothers and the rest of my family. I would like to thank all my lab colleagues at Florida International University (FIU) for their friendship and support, Chandrasekar Puligari, Jorge Blanch, Meng Shi, Andre Senior, Vishnu Madadi, and Adrian Arbide. I also would like to thank Seckin Gokaltun for his generous help and friendship that he offered during my studies. Most importantly, I would like to extend special thanks to Dr. Sabri Tosunoglu for his guidance, patience, and support throughout my years at FIU.

TABLE OF CONTENTS

LIST OF TABLES

LIST OF FIGURES

CHAPTER I

INTRODUCTION

Although humans are highly dexterous and capable, many task requirements well exceed their capabilities. Humans have always been interested in building tools to help them accomplish missions ranging from hunting to making music even in early ages. Simple tool developments have led into the creation of mechanisms in medieval times. As technology progressed and the needs became more sophisticated, simple mechanisms that were devoted to limited tasks proved to be insufficient. New, more complicated and dexterous mechanisms with certain intelligence were needed for use in a wider range of applications. Probably the breakthrough in the development of intelligent mechanisms was the emergence of microprocessors. These small components acted as (or preprogrammed to be) decision makers to relate the actions of the mechanisms to the changes in its environment. Today, microprocessors and mechanisms are used to develop special devices that are capable of performing complicated and precise tasks. These devices that span a wide range in design and capability are collectively called robots.

The term robot was first introduced by the famous Czech writer Karel Capek in one of his plays called Rossum's Universal Robots (R.U.R.) in 1920 [1]. The term "robota" means compulsory labor in Czech language. This definition can be interpreted as robots are devices that obey and accomplish the demands organized by humans.

In 1940's, Isaac Asimov had foreseen the possible progress in robotic research towards artificially intelligent robots that can take initiatives. He also concluded that these initiatives should not be left open-ended. Therefore, he introduced the "three Laws of Robotics" in his science fiction work to restrict the actions of artificially intelligent robots [2].

In fact, the progress in robotics in later years has been towards producing various mechanisms that are artificially intelligent. Today, the term robot has a wider definition. A robot is understood either as an electromechanical device dedicated to accomplishing physical tasks or a software agent used in virtual tasks. The robots in the first category are considered in this study. A more detailed description of the first type of robots is a machine or device that operates automatically or by remote control while interacting with its environment. This denotes the most critical aspects of the robots that they (1) have to be machines or devices, (2) have to be operated autonomously or by remote control, and (3)

1

have to interact with their environments by means of sensors and actuators. Sensors are the key components to build artificially intelligent systems. The system has to first estimate its environment and plan its actions based on this gathered information. Actuators are then used to execute the planned task.

Robots are currently used in all areas of life in many critical tasks. These tasks either require high precision and repeatability which cannot be achieved by humans or tasks that take place in hazardous environments or in unreachable sites by humans. For instance, microscopic surgeries and nano-scale manipulations are examples to the tasks that require high precision. Industrial robots are successfully used in automotive industry because of their high repeatability and cost effectiveness. Space explorations, radioactive material and landmine disposals are some of the tasks that take place in hazardous environments for humans. Mars missions or undersea explorations are categorized as tasks that are performed in unreachable places by humans.

The robots utilized in these applications are either preprogrammed or actively controlled by an operator. Microprocessors are usually used in both types of robots as the decision maker utilizing the gathered sensory information. The preprogrammed robots are often called autonomous robots. The control engineer deploys an algorithm in the microprocessor to evaluate the sensory information and plan the actions without requiring operator's input during the task. Finally, the actuators are driven with the output from the microprocessor to accomplish the action plan.

When an operator's inputs are included in an autonomous robot's control scheme, it becomes an actively operator-controlled robot and generally referred as a telerobot. This type of robotics application is called teleoperation. A teleoperation system consists of a master subsystem, slave subsystem, and a communications line to provide interaction between the two subsystems. Slave subsystem is placed in the task space, which is usually either hazardous or an unreachable site for humans. Master subsystem is located at the same site with the human operator that utilizes it as an interface to control the slave subsystem by issuing demands. The architecture of a teleoperation system and teleoperation application examples are presented in the next chapter.

1.1. Objective of the Work

Teleoperation systems are typically employed in very critical tasks. They are expected to accomplish their missions under extraordinary conditions. These conditions can be classified as variable time delays in communications lines, communication loss, use of different robotic systems, component failures and changes in the system parameters during task execution. The tasks are mostly executed at sites that are unreachable and a failure in the system cannot be fixed by the operator. Such a system working in these conditions should be stable, dependable, and fault-tolerant. The objective of this work is to address the path to developing a teleoperation system of this type.

The research problems that are addressed in this book are presented in three separate areas: (1) development of the teleoperation system model, (2) adaptive control architecture, and (3) development of the fault-tolerant teleoperation system.

The first research area includes development of teleoperation system models for simulation studies and real-time experiments. A new method, Virtual Rapid Robot Prototyping, is used to create these models as accurately and rapidly as possible. Teleoperation system models are configured for a number of different master and slave systems. Master systems include a joystick with fault-tolerant design that enables continuous operation even when an actuator fails, and a commercial joystick. Slave systems range from a mobile platform with fault tolerance features, such as multiple sensors and redundant mechanism design, to commercially available industrial arms.

After the models are created, they are used in simulation studies. The simulations are carried out to test fault-tolerant and adaptive controllers. The necessity of these controllers is explained in the following paragraphs. Later, actual master systems are integrated with the slave systems to perform real-time experiments. The experiments are conducted to verify the simulation results as well as the performance of the controllers.

While modeling the master and slave for the teleoperation systems, a communication line connecting the master and slave robots is also modeled. There is always a communications line involved in a teleoperation system since the local controller (master robot) and the remote system (slave robot) are required to be connected at all times. The information flow from one robotic system to the other can be relayed through various media including the Internet, intranet, satellite or radio signals. The common shortcoming of these communication systems is that they make the teleoperation system experience more significant time delays as the distance between the controller and the remote system increases.

In the past ten years, majority of the teleoperation research has focused on the stability problem that arises under time delays. A control algorithm based on the wave variable technique has provided an acceptable solution for this problem. This algorithm

3

basically stabilizes the teleoperation system with constant time delays in the communications lines.

In contrast, delays in real-time systems are mostly variable time delays. The customary wave variable technique is not always sufficient to overcome instability in the presence of variable time delays. This is where an adaptive algorithm is presented as a suitable solution to ensure stability in this [study].

Adaptive architecture is also required when the system parameters change during the manipulation. The teleoperation system should adapt itself to the changes automatically for stable manipulation. Moreover, teleoperation system is also required to adapt itself automatically when different remote robotic systems are controlled through the same controller. This means that the teleoperation system's operation should not be dependent on the system parameters but be flexible to adapt to any system characteristics, which improves its versatility.

It also becomes necessary to have fault-tolerant architecture in the control scheme when the system experiences communication losses. The reasons for communication losses are that the communications lines are not always dependable and there can always be a disconnection or a faulty component in the system. As a result of fault-tolerant architecture, the task is completed as best as possible, and the damage to the robotic system and its environment is minimized. Position/force controllers are a way of ensuring fault tolerance when the communication is lost in teleoperation.

Another failure is the degradation of the tracking performance during the communication losses. Suitable controllers are required to ensure that the tracking performance is not affected. In addition to these fault tolerant control architectures, all the master and slave systems designed in this work are fault tolerant. This means that even when a component, such as an actuator, fails in any of the systems, the teleoperation system can cope with the failure as best as possible to accomplish the task.

1.2. Contributions to the Literature

A number a new ideas, concepts, and designs are introduced in this work. These form the contributions of this study to the field of robotics, and they are summarized in the following paragraphs.

The first contribution is made in the development of teleoperation system models. A new concept, Virtual Rapid Robot Prototyping, is introduced to create simulations. This method shortens the development time to create simulation models of actual systems while

4

providing a relatively better estimation of the actual system. This is achieved through direct translation of physical and virtual specifications of the actual system from the three-dimensional drawings to the simulation environment. Another approach presented in this ^{work} is a new method to develop virtual haptic environments. The developed virtual laboratory provides an environment to run real-time experiments. The real-time experiments are conducted by using actual controllers to work with virtual robots in the virtual laboratory. Both simulation studies and real-time experiments are conducted to test the adaptive and fault-tolerant controllers.

Adaptive algorithms are successfully applied to teleoperation systems with variable time delays to ensure stability. The effects of the change in wave variables are then observed to adapt to the manipulation speed. As a result of this, the teleoperation system is designed such that the operator input to the master controller automatically generates an appropriate mode for the remote system to run either at a high speed with less accuracy or at a slow speed but with high accuracy.

Another contribution is the implementation of a position/force controller on the slave side to fine-tune manipulations. Since the slave system is always in interaction with the environment that it is working on, it is often a necessity to have control over the force that is exerted on the environment. Alerting the human operator with the help of a force-feedback provided by the slave accomplishes this control in conventional teleoperation systems. Having a position/force controller at the slave side enables the regulation of the force exerted even when the time lag is large. As a result of this, neither the objects the slave robot is working on, nor the slave robot is damaged under excessive amounts of force. This leads to a fault-tolerant teleoperation controller.

The master and slave robot systems are also designed as fault-tolerant systems. Both of the systems are designed to complete the task even when there is component failure in any of the actuators, sensors, and linkage parts. Hence, even if the task completion is not physically possible, the system performance degrades gracefully in the presence of unexpected component failures, and avoids catastrophic outcomes.

The last contribution of this study is to categorize teleoperation systems as limited- and unlimited-workspace teleoperation with respect to their tracking priorities and address the suitable controller for the best tracking performance. The algorithms enable the teleoperation not to lose track even under limited periods of communication losses. This type of controller architecture reinforces the fault tolerance idea in teleoperation.

The result of having these new designs and algorithms is that the limits and boundaries of teleoperation are expanded and the utilization area is widened. Enhanced operation of these systems improves the system reliability and even encourages their use in more critical and diverse applications.

1.3. Outline of the Book

The following chapter provides a review of literature survey on teleoperation systems. The teleoperation concept is then explained by presenting different types of teleoperation system architectures. The concept of fault tolerance in teleoperation is introduced and the types of fault tolerance that can be incorporated in a teleoperation system are explained. Finally, configuration of a teleoperation system is described.

System modeling methods used in this ^{work} is presented in Chapter III. In this chapter, physical model development of a teleoperation system using Computer Aided Design Software and Matlab$^{\copyright}$ is explained. Later, the Virtual Rapid Robot Prototyping (VRRP) idea that combines these two models is introduced.

In Chapter IV, utilizing the VRRP method presented in the previous chapter, development of virtual haptic laboratories is presented. This method is used to create experimental setups for the teleoperation systems tested in this ^{work}.

Fault-tolerant teleoperation components are presented in Chapter V. The master systems, and the slave systems utilized in this study are described in each section of this chapter. The master systems are listed as the (1) two-degree-of-freedom (2-DOF) joystick, and (2) Phantom Omni Device from SensAble Technologies. The slave systems include a variety of robotic devices: (1) 2-DOF joystick, (2) holonomic mobile platform, (3) Epson SCARA robot, (4) Fanuc robot arm, and (5) Motoman UPJ robot arm.

Chapter VI is dedicated to controllers that are utilized in time-delayed teleoperation. In this chapter, the wave variable technique and its modified versions that enhance its usability are presented. Later position/force controllers and their modified versions for teleoperation systems that experience communication loss are explained.

Numerical simulation model development and simulation results are provided in Chapter VII. The first set of simulations is performed using a single-degree-of-freedom teleoperation model whereas the second set utilizes a multi-degree-of-freedom system. The last set of simulations is performed for variable-time delayed teleoperation. The simulation results are presented and conclusions are given.

The results derived from the limited-workspace teleoperation experiments are given in Chapter VIII. Limited-workspace experiments are classified as identical master-slave and redundant slave experiments. The stability and tracking performance of these systems are then examined under constant and variable time delays. Finally, the results of the position/force controllers examined for limited-workspace teleoperation systems are given.

The experimental results are extended for unlimited-workspace teleoperation systems in Chapter IX. These systems are also examined for stability and tracking performances under constant and variable time delays.

In Chapter X, conclusions and findings are listed and recommendations are provided. Finally, the future work is addressed for possible improvements of the teleoperation test system built and the control algorithms that are developed. Also, possible research areas are recommended for future teleoperation studies.

CHAPTER II

LITERATURE REVIEW

2.1. Introduction

Teleoperation represents an application area where humans cannot achieve the job either because the task is too dangerous or needs to be carried out at a distant site from the main control location. Robots, which work in radioactive and hazardous environments, are examples to the robotics tasks that are very dangerous for the humans to accomplish.

Teleoperation systems are composed of a master and a slave subsystem. These subsystems cooperate to complete a task at sites that are either at distant places or at places that are hazardous for humans. The human operator uses a master system to send out commands (or demands). The slave system is driven by these demands. Depending on the information flow, teleoperation systems are usually called unilateral or bilateral. In unilateral teleoperation, no feedback is provided from slave to the master and slave is driven by the commands sent by the master. In bilateral teleoperation, any kind of feedback from the slave to the master can be sent. These feedback signals can be visual, force, sound, position, temperature, radiation, etc. In this study, force-feedback bilateral teleoperation or, in short, force-feedback teleoperation system design, control and implementation is addressed.

In force-feedback teleoperation, as it was the case in unilateral teleoperation, the slave system is driven by the commands sent from the master system. However, in this case, the slave system sends back the force-feedback information that it produces while interacting with the environment to the master system so that the operator has a feeling of the remote system's environment.

In this chapter, examples to teleoperation systems are presented that are found in the literature. Then teleoperation systems are grouped as unilateral and bilateral. Furthermore, bilateral teleoperation is investigated for limited- and unlimited-workspace slave robots. Later, fault tolerance concept in robotics systems is discussed as it also applies to the teleoperation systems. Finally, the concepts presented are outlined in a teleoperation system configuration flowchart.

2.2. Overview of Teleoperation Systems

The goal of teleoperation systems is provide human operators with the ability to interact with environments that would otherwise be difficult to access. Robotic interfaces can provide access to environments that are hazardous, remote, or require interactions at a smaller or larger scale. For example, teleoperated nuclear waste handling systems can keep operators at a safe distance from hazardous material. For space applications, teleoperation systems allow for remote control of such activities as satellite capture and repair. This reduces substantially the risk to humans and the costs associated with manned missions.

Japan's National Institute of Advanced Industrial Science & Technology has concentrated on space robot teleoperation technology to achieve effective ground-based control of manual manipulations in orbit [3]. A representation of their system architecture is shown in Figure 2.1. Similar studies are carried out for ground-space teleoperation at Kyoto University of Japan [4]. Recently, another group of researchers also presented their work on ground-based teleoperation of space robots, which could be used in assembly of space stations and Lunar or Mars explorations [5].

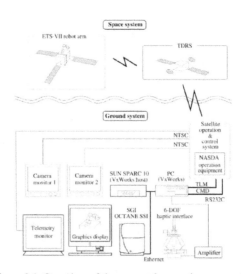

Figure 2.1. Overview of the space teleoperation system [3]

Telesurgery is another developing research field where remote control is required. Cavusoglu at the Univeristy of California at Berkeley has studied telesurgery and surgical simulation [6]. He demonstrated his work using a slave robot at the patient side to tie a knot by the commands sent from the master. Figure 2.2 shows the master and the slave systems used in this study. Also Butner and Ghodoussi at the University of California at Santa Barbara worked on transforming a surgical robot for human telesurgery [7] as shown in

9

Figure 2.3. Other work in this field includes teleoperation of a six degree-of-freedom arm in telesurgery applications [8] and teleoperation of steerable needles for medical use [9].

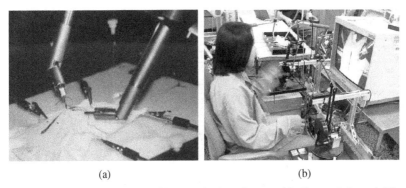

(a)   (b)

Figure 2.2. Master (a) and slave (b) systems tying a knot used in Cavusoglu's work [6]

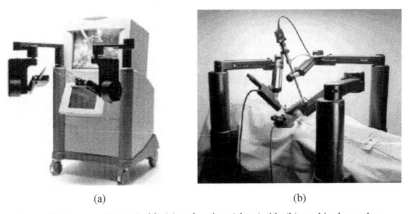

(a)   (b)

Figure 2.3. Surgeon (master) side (a) and patient (slave) side (b) used in the study at University of California at Santa Barbara [7]

Sitti at Carnegie Mellon University has studied teleoperated nanomanipulation [10]. A nano-scale slave manipulator working on nano-particles is controlled with a larger master system. A simple sketch of the test setup is presented in Figure 2.4. These studies are good examples for the wide variety of teleoperation application areas.

Nevertheless, few systems have progressed beyond the laboratory setting. Current telemanipulation systems rely heavily on visual feedback and experienced operators. For example, telemanipulators attached to remote-operated undersea vehicles are position-controlled. They use a miniature replica of the slave arm and they have a visual feedback [11].

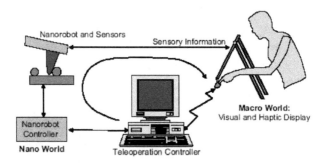

Figure 2.4. Teleoperation control for nanomanipulation [10]

Jet Propulsion Laboratory (JPL) at California Institute of Technology produced a commercial teleoperation system. The slave robot is a mobile platform named Urbie Rover as shown in Figure 2.5 [12]. It has a wide selection of range sensors, a GPS, an omni-directional camera, and binocular stereo camera pair to navigate in urban environments.

Nowadays, military also utilizes telerobots in critical tasks. TALON® Explosive Ordnance Disposal (EOD) robots [13] shown in Figure 2.6 have been in use by the United States military since 2000. They were first used in Bosnia for the disposal of live grenades. Later, they participated in the search and recovery efforts at Ground Zero after the September 11 attack on World Trade Center in New York City, USA. Up to this date, they completed more than 20,000 EOD missions including the military missions in Iraq and Afghanistan.

The quality of a teleoperation experience is often referred to as "telepresence." Ideally, the information from the remote environment (visual, aural, haptic, etc.) is displayed in such a way that the operator "feels" as if he/she is actually present at the remote environment. Presumably, with a greater level of telepresence, an untrained operator can perform tasks as easily as if he/she were at the remote location. The appropriate level of telepresence required for satisfactory performance is still an area of ongoing research.

The first telemanipulation systems, developed in the mid 1940s, provided a direct physical connection with the "remote" environment through mechanical linkages. In 1950s, the linkage connections were replaced with electric servomotors allowing for a much greater distance between master and slave system [14].

11

Figure 2.5 Urbie Rover by JPL [12]

Figure 2.6. TALON® EOD robot [13]

Teleoperation systems that provide force feedback have the potential to greatly enhance the level of telepresence. Research has shown that providing the operator with force feedback can improve task performance [15, 16]. Hannaford et al. [17] showed that during a peg-in-hole assembly task, task completion times, errors, and applied forces were reduced with the addition of force feedback. For these tasks, a force reflecting hand controller was used to control a six degree-of-freedom manipulator with a gripper end-effector.

Researchers in general agree that having force reflection accompanied with visual feedback provides sufficient telepresence for most of the teleoperation applications [18, 19]. Although the addition of force feedback can improve the level of telepresence in a telemanipulation system, several issues are introduced by the inclusion of the operator in the overall feedback loop. When forces are fed back to the operator based on the slave motions and/or interactions between the slave and the environment, the operator becomes dynamically coupled to the master-slave system because the applied forces can now affect the operator's position. This dynamic coupling means that the telemanipulation system is subject to the same constraints of typical feedback control systems. Mechanical compliance, time delays, model inaccuracies, and unaccounted for nonlinearities can all limit the achievable performance and affect system stability.

Anderson and Spong [20] were the first ones to introduce scattering transforms to overcome instability due to time delays. Then Niemeyer and Slotine [21, 22] further developed this concept into the wave variable technique. Munir and Book [23-25] widened the application area of wave variable technique by modifying the existing algorithm with an adaptive algorithm in order to achieve stability even under varying time delays. Hirsche and Buss cooperating with Chopra and Spong [26-29] also worked on a modified wave variable technique to ensure stability under varying time delays. The test systems for these scientists that worked on the wave variable technique is shown in Figures 2.7, 2.8, and 2.9 respectively.

Figure 2.7. The 2-DOF parallel link haptic robot, acting as the master
used in Munir's work [23]

13

Figure 2.8. Master and Slave systems used in Niemeyer's work [21]

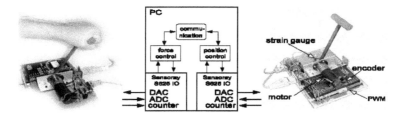

Figure 2.9. Single DOF Teleoperation system used in Hirsche's work [26]

Teleoperation systems are generally classified as unilateral and bilateral depending on the information flow. If the direction of the information flow is only from the master to the slave, this type of teleoperation is titled as unilateral. In bilateral teleoperation information flow is in both ways. The next subsections describe these two groups of teleoperation systems.

2.2.1 Unilateral Teleoperation

In unilateral teleoperation, the information flow is in one direction, or in short, unidirectional. Master system that is driven by the human operator sends the necessary inputs (e.g., position, and/or velocity) through the communications line to drive the slave system. No information is sent back to the master system or the human operator during this type of manipulation. Instead, in most of the cases, the slave system has a local closed-loop control

14

system which uses the feedback signals within this control system. The basic block diagram of unilateral teleoperation is shown in Figure 2.10.

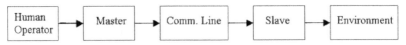

Figure 2.10. Basic blocks of a unilateral teleoperation system

There seems to be a lack of applications of this type of teleoperation in the literature. The reason for this might be that this type of control is not significantly different than any classical control application. Although the human operator sends the command signals to the slave system, actually all the monitoring is accomplished in the control system of the slave.

2.2.2 Bilateral Teleoperation

The basic blocks of a bilateral teleoperation system consist of a human operator interacting with the environment through a slave system as presented in Figure 2.11. The teleoperation system consists of a master and a slave with the communication link between them. Master devices vary from a one degree-of-freedom joystick to glove-based interfaces with many DOF [29]. The slave robotic device may vary from a one-degree-of-freedom "manipulator" to a complex system with a dexterous robot hand attached to a multi degree-of-freedom arm even to multiple slave robots [30]. Both sides of the teleoperation system typically have some type of local control operating on position, velocity, and/or force. The master and slave systems may be controlled by the same computer or by dedicated computers separated by thousands of miles.

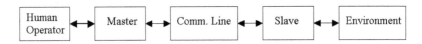

Figure 2.11. Basic blocks of a bilateral teleoperation system

A common type of teleoperation architecture is one in which the master system sends position or velocity commands to a slave system. Force or torque information induced from interactions with the environment is fed back to the master system (by the slave system). This type of two-channel (one communication link in each direction) architecture, as depicted in

Figure 2.12, is often referred to as "position-force" architecture. Another type is the four-channel teleoperation, as shown in Figure 2.13, where both positions (or velocities) and forces are transmitted between the master and slave [31].

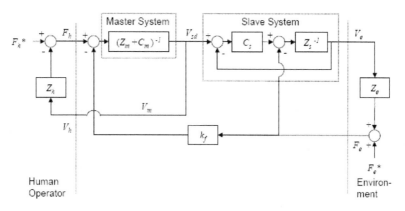

Figure 2.12. Two-channel architecture for bilateral teleoperation [31]

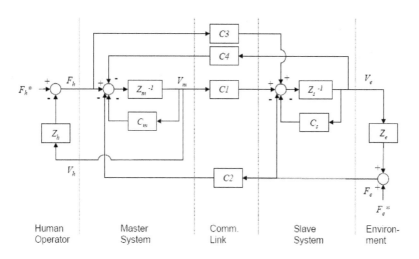

Figure 2.13. Four-channel architecture for bilateral teleoperation [31]

It appears that some researchers chose to work with the four-channel model [31-34] while others chose to work with the two-channel model because the information flow was

16

sufficient enough for a stable teleoperation. These researchers also introduced the wave variable technique for stability even under time delays [20-28]. This technique is described in Chapter VI. Aziminejad et al. worked with both bilateral teleoperation architectures in their effort to apply the wave variable technique to both architectures [35].

In this study, bilateral teleoperation systems are further investigated in two groups as limited- and unlimited-workspace teleoperation. These two groups are introduced in the following subsections.

2.2.2.1 Limited-Workspace Teleoperation

Teleoperation systems using serial or parallel slave manipulators with limited workspace are defined as limited-workspace teleoperation. Telemanipulation of an industrial robot arm is a typical example to this type of teleoperation. Researchers working on the stabilization issue of the time-delayed teleoperation have also chosen to work with limited-workspace teleoperators [22, 24, 27].

In limited-workspace teleoperation, generally master position and orientation information is mapped into the Cartesian position and orientation of the end-effector. Therefore, position tracking becomes a priority for the limited-workspace.

2.2.2.2 Unlimited-Workspace Teleoperation

Teleoperation systems composed of a mobile platform or any unlimited-workspace slave is referred as unlimited-workspace teleoperation. Telemanipulation of any mobile robotic system whether it operates on ground, water or in air is categorized as unlimited-workspace teleoperation. For instance, researchers at Santa Clara University have worked with unlimited-workspace teleoperators in their studies [36]. Figure 2.14 shows the teleoperated Triton Undersea Robot used in these studies. Some other unlimited-workspace systems that they have used are the Roverwerx Terrestrial Robot, and the Bronco Unmanned Aerial Vehicle (UAV).

JPL has worked in teleoperation field and produced various teleoperation systems. One of these systems is the Urbie Rover that was explained in the previous section. This system is another example to the unlimited-workspace teleoperation.

TALON® EOD robot is another example to the unlimited-workspace teleoperation. It is a mobile platform with a robotic arm placed on top of the platform. It has been used in numerous critical military tasks in years.

17

Figure 2.14. Teleoperated Triton Undersea Robot [36]

In unlimited-workspace teleoperation, the position information from the master is generally mapped as velocity demand for the end-effector of the slave. The tracking priority is given to the velocity in unlimited-workspace teleoperations.

2.3. Fault Tolerance in Teleoperation

A fault tolerant system is one that can identify a failure, isolate that failure and provide a means of recovery. Teleoperation robots are designated to work in deep-sea operations, space missions, nuclear cleanup, and any remote operations. Because these robots are in situations that are hazardous for humans or remote from the human operators, a robot failure can be very expensive. In these critical missions, robotic systems must be fault tolerant.

Fault tolerance is increasingly seen as an important feature in modern autonomous or industrial robots. The ability to detect and tolerate failures allows robots to effectively cope with internal failures and continue performing designated tasks without the need for immediate human intervention. This is crucial in teleoperation applications because the slave system is usually unreachable to the human operator to be fixed at a time of failure.

In the past, fault tolerant systems have been developed for computer, aerial vehicles, and industrial systems [37-39]. Some of these techniques utilized models for robotic fault tolerance schemes [40]. Robotics engineers rely on physical redundancy of components when applying fault tolerance to their systems [41]. Robotic fault tolerance is categorized as computer and sensor fault tolerance, actuator and joint fault tolerance, and mechanism fault-tolerance.

2.3.1 Computer and Sensor Fault Tolerance

Triple Modular Redundancy (TMR) is usually used to provide fault tolerance in computer systems or sensor systems [42]. In this method, three processors or sensors work on the same problem and compare the outcomes shown in Figure 2.15. When one of the processors or sensors is faulty and its outcome does not agree with the results of the other two processors or sensors, the faulty processor or sensor is voted out of the final decision. A shortcoming of this process is that only one faulty processor or sensor can be tolerated by this system.

The TMR voting scheme can be expanded by adding more redundancy in terms of processors or sensors. NASA Space Shuttle uses five redundant General Purpose Computers in an expanded TMR voting scheme [43]. There are four computers that are exact duplicates and work redundantly to perform the same tasks given the same input data. The four computers vote on the results and can detect up to two flight-critical computer failures. If two failures happen, the two computers remaining use comparison and self-test methods to tolerate a third failure [43]. Backup Flight Software by the fifth computer and it does not generally perform critical flight functions. The fifth computer could be used as a backup if the failure is due to an architectural design flaw in the main processors [44].

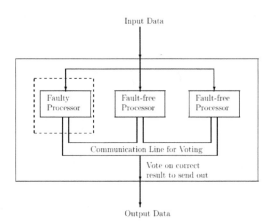

Figure 2.15. Triple Modular Redundancy [42]

2.3.2 Actuator and Joint Fault Tolerance

Fault tolerance for the mechanical failures of robots previously has concentrated on algorithms that rely on duplicated parts. Generally, these schemes concentrate on faults in a specific part of the robot such as a motor, sensor or joint failure. Robotics researchers have worked on duplicating actuators in a robot joint [40, 45]. In their fault-tolerant scheme, the two actuators in a joint must be able to work together to provide one output velocity for the joint. Therefore, if one of the motors breaks, the other one takes over the faulty motor's functions while compensating for any transients as a result of the failed motor. As a result of this, for a robot that is performing a time-critical or delicate task, this type of fault tolerance allows the robot to get a run-away actuator under control quickly before environment or the robot damaged. This is vital in teleoperation cases where the slave manipulator is at a remote and unreachable site carrying out critical tasks.

Redundancy, providing the advantage of fault tolerance, has also led to adding extra parallel structures, such as a backup arm or leg [40]. This allows different reconfiguration possibilities when failure occurs. Redundant components are utilized in the reconfiguration problem as they provide a backup. In order to improve fault detection, redundancy can also be used to give the robot system multiple components to check and vote between.

2.3.3 Mechanism Fault Tolerance

Today many robots are built with the advantage of being kinematically redundant. Kinematic redundancy means that the robot has more degrees-of-freedom or motions than necessary to position and orient the end-effector which allows the robot to choose between multiple joint configurations for a given end-effector position in the robot workspace. The kinematic redundancy is used in developing fault-tolerant algorithms that use the alternate configurations in positioning and orienting a robot with failed joints. These algorithms do not require extra motors, sensors or other components to be added to the robot. They use the existing structure to provide fault tolerance within the existing physical limitations [46].

Maciejewski studied the effect of joint failure on the dexterity of a kinematically redundant manipulator [47]. He calculated an optimal initial configuration to maximize the fault tolerance of redundant arms while minimizing the degradation of the system performance during a failure in the system.

An aspect of fault tolerance that should be taken into account occurs when a robot is in a singular configuration, which may trigger a false alarm for failure. When robot is fully extended or folded in on itself in such a way as to hinder motion in one direction without rapid changes in one or more joint positions, this type of configuration is called singular. When the robot passes through one of these singularities, the joint velocities of a manipulator becomes extremely high. These extreme joint velocities can be interpreted as failures in the robot and erroneously shut down a fault-free system by the fault-detection algorithms. The optimal damped least-squares technique used in the Singularity Robust Inverse (SRI) algorithm provides feasible joint velocities which results in minimal deviations from the specified trajectory of the end-effector [48]. As a result of this, SRI helps the manipulator to avoid drastic joint motions at or near singular configurations and eliminates false alarms for fault detection.

2.3.4 Control Algorithms for Fault Tolerance

Fault-tolerant control of a robotic system under the failure of its components is not an easy task. The goal is to have a robust control over the system with a minimum recovery time in the face of disturbances caused by the component failures so that the system continues to work on the given task [49]. Researchers examined various control algorithms for fault-tolerant control. These control algorithms can be listed as computed-torque method [50], sliding mode controller [51], adaptive [52], and fuzzy logic controller [53]. A comparison study for these controls is presented in Monteverde and Tosunoglu's paper [54].

The dynamics of an n-DOF robotic system can be given by,

$$\tau = M(q)\ddot{q} + C(q,\dot{q}) + G(q) + F(q,\dot{q}) \tag{2.1}$$

where τ is the n input actuator torques, $M(q)$ is the nxn generalized inertia matrix, q is the n-dimensional joint displacement vector, $C(q,\dot{q})$ is the centrifugal and Coriolis terms, $G(q)$ is the gravitational terms, and $F(q,\dot{q})$ is the frictional load terms.

The computed torque method can be formulated using the given dynamics above as,

$$\tau = \hat{M}(q)(\ddot{q}_d + u) + \hat{C}(q,\dot{q}) + \hat{G}(q) + \hat{F}(q,\dot{q}) \tag{2.2}$$

where u is the feedback control law and it can be described for the PID control as,

$$u = K_v \dot{e} + K_p e + K_I \int e \, dt \tag{2.3}$$

where $\hat{M}(q), \hat{C}(q,\dot{q}), \hat{G}(q)$ and $\hat{F}(q,\dot{q})$ is the online calculated values. In the above equation, $e = q_d - q$ is the position error vector, $\dot{e} = \dot{q}_d - \dot{q}$ is the velocity error vector and K_p, K_d, K_I are the gain matrices.

Sliding control is formed from the variable system and tries to take the errors into a sliding surface to make the tracking errors approach to zero. Sliding mode control is known to be more robust than the computed-torque method, which makes it more reliable under the component failures can cause instabilities to be dealt with. A sliding surface is to be selected as a function of tracking errors,

$$S = \dot{e} + \lambda_1 e + \lambda_2 \int edt \tag{2.4}$$

where λ_1 and λ_2 specify the response of the error dynamics in terms of the bandwidth or percentage of overshoot.

Adaptive control with parameter estimation tries to identify the uncertain parameters of the system in real-time. Because there are always uncertainties in robotic systems, this is a very useful method to overcome the uncertainties in fault-tolerant systems. Generally, the unknown parameter is identified by re-parameterizing the dynamic model used in the controller.

Fuzzy logic control utilizes IF-THEN rule base to drive the output. It also uses membership functions to provide the continuity of control inputs rather than the on/off Boolean logic strategy.

2.4. Teleoperation System Configuration

The path to develop a teleoperation system is summarized in the flowchart presented in Figure 2.16. This chart provides guidance to the design engineer to decide on the critical design parameters of a teleoperation system so that the system to be developed serves its intended purpose [55].

A design engineer first has to decide whether the application requires fault tolerance or not. If the task must be completed under a possible component failure, or requires higher reliability specifications, the decision should be definitely towards a fault-tolerant design. System architecture should be configured considering this decision. The next step is to select fault tolerance components to be used in the design.

22

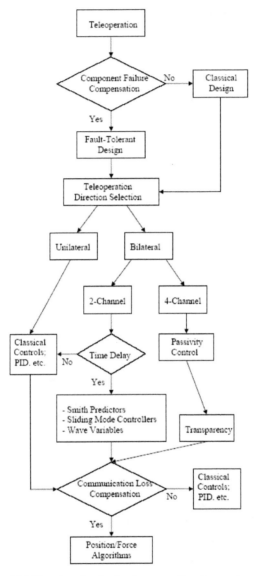

Figure 2.16. Flowchart for the development of a teleoperation system

The designer then selects the type of information flow between the master and the slave. He may either choose to monitor, or not to monitor the slave manipulation. By selecting a unilateral teleoperation, he chooses not to monitor; hence, the slave actions are solely monitored by its built-in control architecture. In other words, the slave runs in an open-loop fashion. Selecting bilateral teleoperation means that the actions of the slave robot are monitored by the feedback signals provided by it. The designer has to decide the quantity of feedback signals as well. He may select to have one feedback signal by selecting 2-channel, and two feedback signals by selecting 4-channel teleoperation.

Later the designer is expected to decide whether he wants to have control over time delays that develop in the communications line or not. He may select a wave variable controller, sliding mode controller, or another controller to have a stable teleoperation even under time delays. These controllers are further explained in Chapter V.

The last step of configuring a teleoperation system will require the designer to decide whether or not position/force compensation will be activated on the slave side in case a communication loss occurs between the slave and master systems. If the compensation is desired, then the appropriate position/force control algorithm needs to be selected among a menu of available algorithms. The most common position/force controllers are also presented in Chapter VI.

CHAPTER III

VIRTUAL RAPID ROBOT PROTOTYPING METHOD

3.1. Introduction

In this chapter, modeling and simulation of teleoperation systems is addressed by introducing the concept of virtual rapid prototyping. Rapid prototyping is one solution to produce a design in minimum amount of time, which improves efficiency and substantially reduces costs.

Development of the robot design starts with a Computer-Aided Design (CAD) model, system assembly, and simulation of the system motion. It then progresses to the development of kinematic and dynamic model simulations, structural design, and is completed by designing a controller for the robot. This streamlined approach allows easy reiteration of the design process at any stage; thus, it allows the designer to optimize system parameters as much as possible. The rapid prototyping environment presented in this chapter is developed by integrating SolidWorks$^©$, Matlab$^©$ and their modules to create a detailed system model for use in simulations and controller development. Although the process is applicable to the design of any mechanical system, robots with their high degrees of freedom are especially suitable for virtual rapid prototyping.

CAD software is a tool to design any physical system either in two- or in three-dimensional space in a virtual environment. This tool is especially useful for designers in the design of multi-degree-of-freedom (DOF) robots to test their performance even before manufacturing them. As professional software packages evolve, more functions are made available to the design engineers. However, to design a robotic system, usually several software tools are needed. This work addresses the integration of two software packages; namely, SolidWorks$^©$ and Matlab$^©$, to design robotic systems.

SolidWorks$^©$ is a powerful CAD tool to design system parts and assemblies, and animate the system motion utilizing its animation tool CosmosMotion. For certain types of parallel mechanisms, the software performs kinematic and dynamic analysis while the robot is in motion to calculate various physical quantities of the system such as the forces exerted on joints, positions, velocities, accelerations, and so on. The capability of the software can be extended to robots that use serial architectures by carrying out the kinematic analysis of the robot externally and then importing the data to SolidWorks$^©$ for animation purposes. There are other kinematics solution packages that can be used in mechanism design and one example is SAM$^©$ software.

Another helpful function of the software is also addressed by transferring the robot's data into the Matlab$^©$. The visual representation of the robot is transferred to Matlab$^©$ Virtual Reality environment as VRML files. This enables the designer to view the robot in action while the simulation is running.

In order to create simulations, Matlab$^©$ introduces the Simulink environment and Simmechanics blocks which can accomplish forward kinematics and dynamics modeling. Control algorithm of the robot can also be developed in this environment. Matlab$^©$ also released a new translator to translate SolidWorks$^©$ forward kinematics and dynamics information into Matlab$^©$ as Simmechanics blocks [56]. The blocks created are explained in the following sections.

Final sections of this chapter are dedicated to describe the integration of Virtual Reality model with Simmechanics model. Robot prototyping samples utilizing SolidWorks$^©$ and Matlab$^©$ are also provided.

3.2. Animation and Analysis Background

Today's CAD tools cannot be considered simple software tools for two- or three-dimensional sketching since their capabilities allow them to be utilized as analyzers as well. The finite element analysis is probably the most-commonly used tool of many analysis tools that are available for these CAD tools. In this work, the focus is on robot design, its animation, and eventually its kinematic and dynamic simulations with the intent to develop and implement controllers for the designed system. To accomplish this goal, the initial aim is to develop a robot using one specific CAD tool; SolidWorks$^©$, and then animate the robot in the same CAD environment.

SolidWorks$^©$ is one of the capable CAD tools currently available besides Pro-Engineer$^©$, AutoCad$^©$, Unigraphics$^©$, I-Deas$^©$ and others. Most of these programs have their own animation tools and they can also output VRML files to develop Virtual Reality environments. The significance of SolidWorks is that Matlab$^©$ recently introduced a tool to work with SolidWorks$^©$ so that physical properties are extracted from the SolidWorks$^©$ model and used automatically to create Matlab$^©$ Simulink models. The physical properties extracted include mass, inertia, center of gravity, position and orientation, link length and other geometric parameters.

Although most of the CAD tools have mechanism design and animation capabilities, each has different procedures. The procedure described in this chapter is for SolidWorks[©] CosmosMotion mechanism design tool.

SolidWorks[©] is utilized for product design purposes by many companies that range from aerospace and defense to automotive industries. For instance, Alliance Spacesystems, Inc. used SolidWorks[©] to develop robotic arms for NASA's Mars Exploration Rover (MER) mission [57]. After using the stress and thermal analysis tools to optimize the design, the system is further analyzed by animating the mechanism in the animation module. Figure 3.1 presents the photo of MER accompanied with the SolidWorks[©] model of MER's robotic arm.

(a)   (b)

Figure 3.1. (a) MER; (b) SolidWorks[©] model of MER's robotic arm [57]

The U.S. Army Research Laboratory also develops their designs for military applications in SolidWorks[©], and later uses ANSYS and CircuitWorks in the analysis stage [58]. In the automotive industry, Commuter Cars Corporation [59] manufactures world's fastest urban car Tango after they have configured their designs in SolidWorks[©]. Figure 3.2 shows the urban-commuting vehicle, Tango, and its structural CAD design.

Currently, many robotics researchers use these CAD tools in their simulation works. For instance, Pap, Xu and Bronlund study kinematic simulations of a robotic human masticatory system using CosmosMotion [60]. Some researchers at Tokyo-based Speecys Corp. developed their humanoid robots and their gaits using the SolidWorks[©] CosmosMotion environment [61].

Robotics researchers at Florida International University (FIU) also chose to create their gaits using SolidWorks[©] and CosmosMotion [62]. After creating the parts and assemblies for the robot, they also configured it as a mechanism to animate it for testing new

27

gaits. While testing the gaits before building the robot, they also analyzed the design for optimization purposes.

(a)   (b)

Figure 3.2. (a) Tango; (b) Tango's structural design in CAD environment [59]

In another humanoid study at FIU, researchers developed a kinematic simulation environment for humanoid studies [63]. The walking gaits of the humanoid are first created and then simulated in the program. Later, the source code to be embedded in the microprocessor is created automatically by this program. The main interface window of the program is shown in Figure 3.3.

Nowadays, a number of dynamics and kinematics simulation tools are commercially available such as Adams$^{©}$, Labview$^{©}$ and Matlab$^{©}$. Among these, Matlab$^{©}$ is probably the most used tool for robotics researchers as it offers two separate platforms to create simulations. The matrix-based programming language of Matlab$^{©}$, M-file, makes the software a popular choice for robotics engineers that perform matrix-based solutions for robot kinematics and dynamics. Matlab$^{©}$ also offers a large cluster of subroutines in its block-based simulation environment, Simulink. These subroutines are called blocks and they are wired together to form simulations. Special purpose blocks are collected together at smaller clusters. One of these clusters is the Simmechanics toolbox, where Matlab$^{©}$ offers blocks to configure mechanisms and robotic devices. Simmechanics blocks include a wide range of robot components such as actuators, sensors, bodies and joints.

Today, majority of the robotics simulation studies presented in research articles are created in the Matlab$^{©}$ Simulink environment [56]. Similarly, companies introducing cutting-edge technologies also heavily rely on Matlab$^{©}$ in creating simulations to test their algorithms and products. For instance, Boeing engineers used Matlab$^{©}$ Simulink environment to create a system model and simulation-test the flight control laws for their X-40A Space Maneuver Vehicle (SMV) [64]. Daimler Corp. also designed, tested and implemented the cruise controller for the Mercedes Benz trucks using the toolboxes offered by Matlab© [65]. Similarly, Lockheed Martin Space Systems utilized Matlab© to configure their control designs and automate the development of accurate, real-time spacecraft simulations [66].

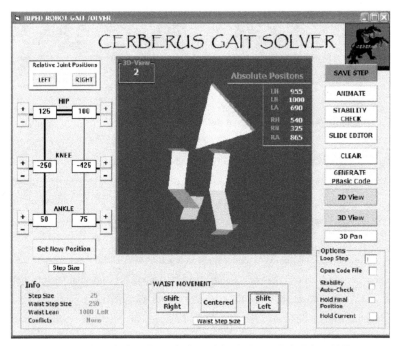

Figure 3.3. Main interface window of the Cerberus Gait Solver [63]

3.3. SolidWorks© Modeling and Animation

Part creation and assembly development are considered relatively straightforward for today's designers. The challenge is now on quick modeling, simulation and animation of these mechanisms. The following paragraphs describe the creation of a mechanism model and animation in SolidWorks©. A two degree-of-freedom (DOF) revolute-prismatic (RP) manipulator is used as a sample in this section.

Assembly creation process becomes very important if the mechanism is to be animated. The mating process should be accomplished carefully. Parts that are screwed, welded or somehow fixed to each other should not have a motion relative to each other. If

29

there is any kind of joint placed between two parts, it is a definite sign that the parts are not fixed relative to each other. On the other hand, when creating a joint, two parts to form the joint should have a motion, either translational or rotational with respect to each other.

CosmosMotion lets the user to input motion to the joint; thus, it can create an animation of the mechanism. The sample mechanism or robot arm we are using to demonstrate the procedure in this chapter is a revolute-prismatic (RP) manipulator. If the mating process is accomplished correctly, the mechanism should have one revolute (R) and one prismatic (P) joint. Figure 3.4 shows the motion-building window of CosmosMotion. As it can be observed from the figure, definitions of the mating conditions result in the creation of one revolute and one prismatic joint.

It can also be observed from Figure 3.4 that the motion for the "Revolute" joint is described as a constant speed of 360 deg/sec or in other words 60 rpm. There are different ways of describing the joint motions. Probably the most efficient way is to calculate the joint trajectories as a result of the inverse kinematics solutions for the given end-effector trajectory and then to input the joint motion as a function of time for each joint.

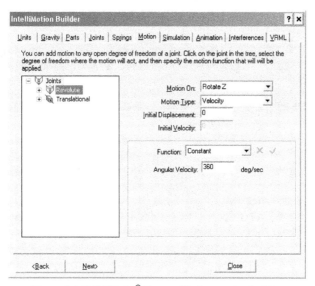

Figure 3.4. SolidWorks©/CosmosMotion motion builder

After simulating the mechanism for the pre-described joint motions, the animation created can also be saved as a movie file. Figure 3.5 shows the window to save the animation movie. "Create Animation" button on the window saves the animation to the designated location on the hard disk.

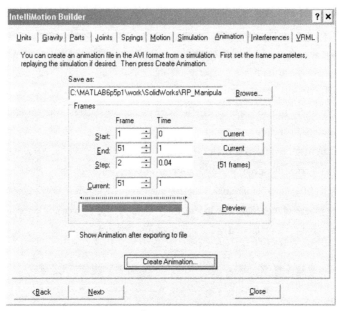

Figure 3.5. SolidWorks©/CosmosMotion animation window

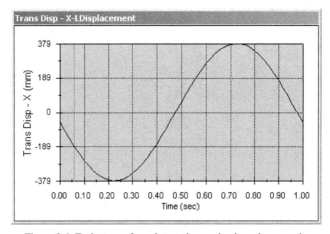

Figure 3.6. Trajectory of a point on the mechanism along x-axis

Following the animation creation, various analyses on the system can be carried out. For instance, one analysis includes path following of any point on the mechanism and its trajectory on each world axis. The vertical solid line at 0.05 second on the trajectory plot example shown in Figure 3.6 can be moved forward and backward synchronized with the mechanisms posture at that time interval. Figure 3.7 shows the main window of SolidWorks© with the path drawn for a certain point of the mechanism and the trajectory of that point along *x*-, *y*-, and *z*-axes.

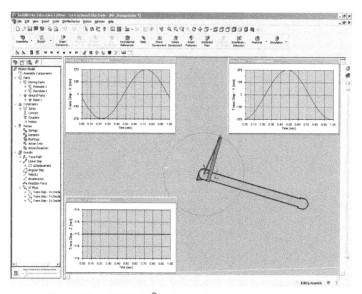

Figure 3.7. SolidWorks© main window with analysis results

3.4. Matlab© Modeling

Matlab© models of physical systems can be created in two different environments that Matlab© provides. A matrix-based programming language called M-files is one Matlab© environment that the programmer can create the model by writing the code line-by-line [56]. Simulink is the other programming platform that Matlab© offers [56]. This platform utilizes a drag-and-drop graphical-user-interface; hence, it is a more user-friendly option for model creation and simulation purposes for the designer.

Simulink coding utilizes previously-defined blocks instead of developing the program line-by-line. As Simulink environment became a major simulation environment for most of the researchers, the platform also evolved in time providing a variety of blocks for modeling purposes. The new branch of blocks that the robotics researchers are interested in is collected in Simmechanics blockset. This blockset offers link, joint, actuator, and sensor blocks to develop the simulation model of any physical system that uses these components. Some of these blocks are introduced in Table 3.1.

Table 3.1. Matlab$^©$ Simmechanics blocks

	"Ground" block grounds one side of a joint block to a fixed location in the World coordinate system
IC	"Joint Initial Condition" block sets the initial linear/angular position and velocity of some or all of the primitives in a joint block.
	"Joint Actuator" block actuates a joint block primitive with generalized force/torque or linear/angular position, velocity, and acceleration motion signals.
	"Joint Sensor" block measures linear/angular position, velocity, acceleration, computed force/torque and/or reaction force/torque of a joint primitive.
	"Revolute" joint block represents one rotational degree of freedom. It can be driven by the "Joint Actuator" block and its motion can be measured by the "Joint Sensor" block if the blocks are attached to this block.
	"Prismatic" joint block represents one translational degree of freedom. It can be driven by the "Joint Actuator" block and its motion can be measured by the "Joint Sensor" block if the blocks are attached to this block.
	"Body" block represents a user-defined rigid body. "Body" block is defined by mass, inertia tensor and coordinate origins.
	"Body Sensor" block measures linear/angular position, velocity, and/or acceleration of a "Body" block with respect to a specified coordinate system.

3.4.1 Creating Simmechanics Model from SolidWorks$^©$ Model

Simmechanics model of a physical system (such as a robot manipulator) can be created from scratch by using the blocks explained previously [56]. The mass, inertia,

33

orientation, and center of gravity information for the links and joints can be input manually when creating the system from scratch. Although this is possible by getting this information from SolidWorks© manually, it is time consuming for complex systems with higher degrees-of-freedom, and prone to introduce errors as link/system parameters are changed during the design phase.

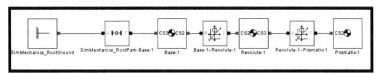

Figure 3.8. Simmechanics model of an RP manipulator translated from SolidWorks©

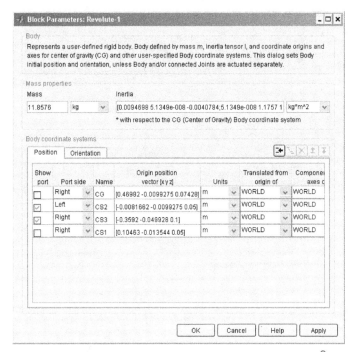

Figure 3.9. Revolute body block parameters translated from SolidWorks© model

Mathworks recently released a translator to translate SolidWorks© models into the Simmechanics environment. The end result of this translation is a forward kinematics and dynamics model of the system created in SolidWorks©. The model is created using

34

Simmechanics blocks and all the necessary information (mass, inertia, orientation, etc.) is automatically transferred to these blocks. Figure 3.8 shows the automatically-created Simmechanics model of an RP robot manipulator.

The Simmechanics blocks that are created as a result of SolidWorks© to Simmechanics translation are illustrated in Figure 3.8. The base part, "Base," of the manipulator is fixed to the ground, which is described as a zero-DOF with respect to the ground. The base is then connected to the revolute link, "Revolute," via a revolute joint, which is described as one rotational DOF in between the base and the revolute link. The revolute link is then connected to the end-effector, "Prismatic," via a prismatic joint, which is described by a translational motion between the revolute link and the end-effector.

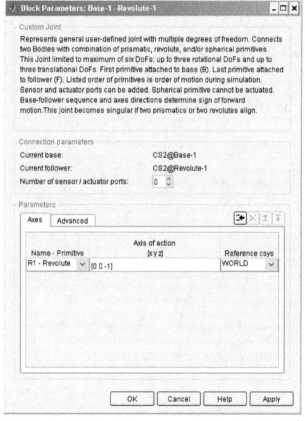

Figure 3.10. Revolute joint block parameters translated from SolidWorks© model

35

Figure 3.9 shows the "Body" block for the revolute link where the mass, inertia and length information is translated from SolidWorks[©]. Figure 3.10 shows the translated revolute joint and the rotation axis information. "Sensor/Actuator" port is to be increased to connect sensors and actuators for simulation purposes in later stages.

3.4.2 Creating Virtual Reality Model from SolidWorks[©] Model

Even in the early design stages, it is advantageous to the design engineers to observe the motions of a robotic system. This would provide them a better visual test of the system before manufacturing the robot, and allow them to redesign some of the parts if necessary after inspecting the animations. Matlab[©] provides this opportunity in two different ways. One is the built-in visualization tool that develops the visual representation of the model automatically. This tool basically draws straight lines from one node to another for each link. It also shows the axis system and Center of Gravity (COG) for each link and joint if these options are activated. Figure 3.11 shows the visual representation of a 6-DOF robot manipulator drawn by using the visualization tool. All the links and joints are synchronized with the model, which means that as the simulation runs, the visual representation of the model updates itself accordingly.

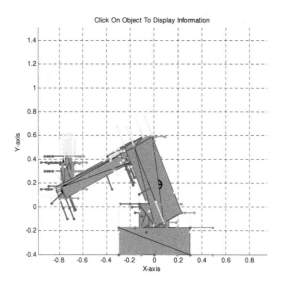

Figure 3.11. Visual representation of a 6-DOF robotic arm by visualization tool

36

Another option to create the visual representation of the Simulink model is to use the Virtual Reality Toolbox. This toolbox enables the user to import the 3D CAD models into the Virtual Reality (VR) screen. The motion of the links and joints are then coordinated using V-Realm Builder and "VR Sink" block of Simulink. Once the coordination is complete, the animation of the 3D model is much faster and relatively smoother visually in Matlab$^©$ than in the SolidWorks$^©$ animation. Figure 3.12 shows the VR representation of the same robotic arm in the previous figure.

Trajectory creation and joint motion creation in Matlab$^©$ simulations are also easier with respect to SolidWorks$^©$ animation creation. In SolidWorks$^©$ animation, an external software module should be used to solve for inverse kinematics of the mechanism to calculate each joint motion. Whereas in Matlab$^©$, inverse kinematics solution can also be carried out within the same simulation.

Figure 3.12. Visual representation of a 6-DOF robotic arm by VR screen

3.4.3 Integration of Virtual Reality Model with Simmechanics Model

Virtual Reality model and the Simmechanics model are both created using the SolidWorks$^©$ model. These two models are required to be integrated and synchronized using Simulink blocks. As described previously, "VR Sink" block is utilized for this purpose. The inputs to this block are to be created from the Simmechanics block for synchronization purposes.

The rotation centers are created as the origin (0,0,0) in the world axis system. This is not true for the links translated from the SolidWorks© model to VRML. The centers of rotation should be corrected by using the information in the "Body" blocks of the Simmechanics model.

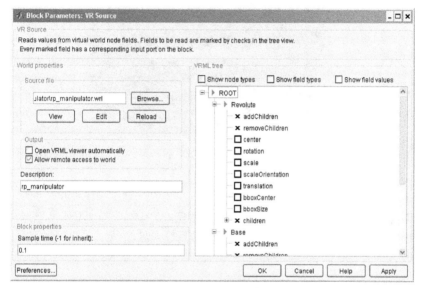

Figure 3.13. VR sink block parameter window

Figure 3.13 shows the "VR Sink" block parameter window when the RP manipulator VRML is loaded. As seen in this figure, there are boxes next to some of the parameters that are left blank (unchecked). The value of these parameters can be provided from the Simmechanics blocks continuously during the simulation. For example, the boxes for the center of rotation and the rotation amount about any axis can be checked, and; therefore, controlled during the simulation to rotate the part about that center in the VR screen.

After all the boxes are checked to represent the motion that the Simmechanics blocks are performing, the "VR Sink" block opens ports to interact with the simulation environment. The VR screen for the RP manipulator is shown in Figure 3.14. Connecting the necessary inputs from the Simmechanics blocks to these ports, the integration of visual representation of the mechanism with the Simmechanics blocks is completed.

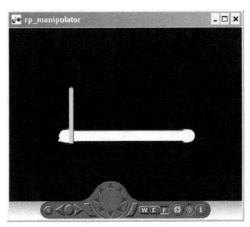

Figure 3.14. Virtual reality representation of RP manipulator translated from SolidWorks[©]

As a last step before running simulations, simulation solver type and the time-step size need to be specified for customized use. Figure 3.15 shows the simulation parameter window. Gravity is also modeled in the environment to represent the outer world accurately and it can be switched on or off as required by the simulation scenario.

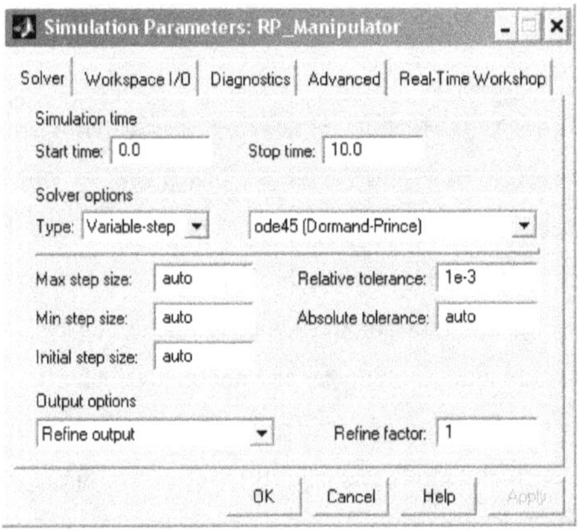

Figure 3.15. Simulation parameters window

3.5. Sample System Models

An example simulation is carried out for the serial RP robot manipulator to demonstrate the procedure outlined above. The RP manipulator in this example is developed as a planar robot. The simulation is carried out in two different phases. The first phase is the kinematics simulations. The reason to start with kinematics simulations is to verify the inverse kinematics solutions as well as the task created. The sample task requires the end-effector to follow a straight-line trajectory.

Figure 3.16 shows the main window of the simulation. In this window, integration of the Simmechanics model with the VR model can be observed from the connections made to the "VR Sink." The end-effector's trajectory is also followed by another scope that appears as the "X-Y Graph" on the bottom right side of the window. Simmechanics blocks are hidden inside the block named "RP_Manipulator" on the top right corner of this window.

The Simmechanics blocks created as a result of the translation from SolidWorks$^{©}$ were presented previously. New Simulink blocks are added to these blocks to actuate the mechanism as shown in Figure 3.17.

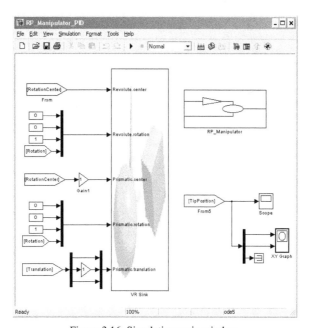

Figure 3.16. Simulation main window

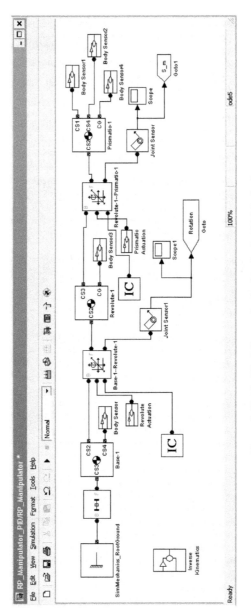

Figure 3.17. Modified Simmechanics blocks for RP Manipulator

Another block added to the basic system model is the inverse kinematics solution. This block is formed using simple Simulink blocks to solve for joint motions when the end-

41

effector motion is specified. Figure 3.18 shows the inverse kinematics calculations for this manipulator.

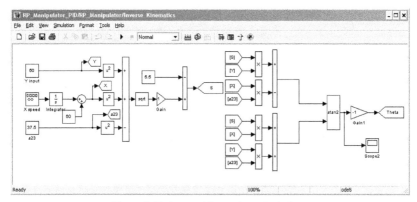

Figure 3.18. Inverse kinematics calculations

Joint actuation for the kinematics solution and dynamics solution differ. In kinematics solution, joints are actuated with the provided joint trajectories as a function of position, velocity and acceleration. Dynamic effects are neglected in this solution. Figure 3.19 shows a sample joint actuation for the kinematics solution. As it can be observed, there is no control law over the motion fed to the actuator. Therefore, in the kinematics solution the dynamic effects do not have influence over the motion and the result comes out to be a perfect straight-line path followed by the end-effector. The line drawn as a result of this simulation is shown in Figure 3.20.

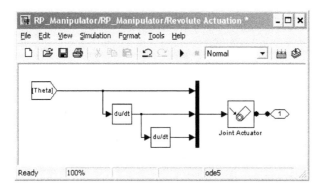

Figure 3.19. Joint actuation for the kinematics solution

42

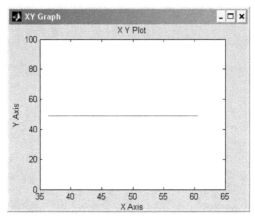

Figure 3.20. End-effector's path for the kinematics simulation

The joint actuation blocks are modified for the simulations taking into account the dynamics of the mechanism. As it can be observed from Figure 3.21, a PID controller is used as the control law for the revolute joint actuation. In addition, compensation for gravitational effects is added to the joint torque calculation. Also "Initial Condition" blocks are used to start the joint actuation close to the initial position. This minimizes transition state overshoots for each joint.

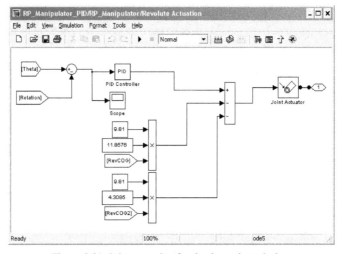

Figure 3.21. Joint actuation for the dynamics solution

43

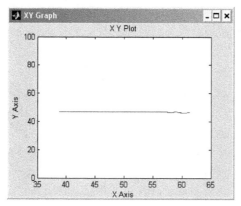

Figure 3.22. End-effector's path for the dynamics simulation

A transition state is expected for every mechanism at the start of manipulation. This transition state can be clearly observed from the path of the end-effector shown in Figure 3.22.

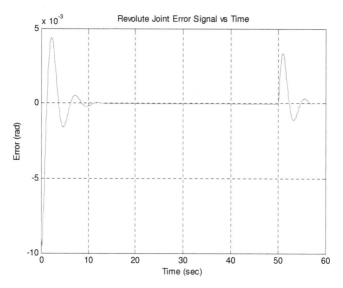

Figure 3.23. Revolute joint error signal

The transition state can also be observed from the error signal created. Figure 3.23 shows the error signal plot for the revolute joint. Looking at these graphs during the manipulation, the designer can either set the control parameters or change the control algorithm. The specific task was performed at a relatively low speed (~0.5 cm/sec). Therefore, the effect of the centrifugal and Coriolis force terms may be neglected in the control design. For faster manipulation speeds, these effects cannot be neglected. Thus, nonlinearity cancellation can be incorporated by employing computed-torque control.

Another simulation study was carried out for the SCARA arm E2H853C by Epson [67] and the VRRP method is used. Figure 3.24 shows the VR model and the actual picture of the robot.

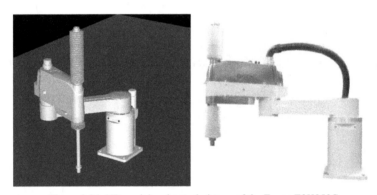

Figure 3.24. VR model and actual picture of the Epson E2H853C

Simulink blocks of the model are shown in Figure 3.25. The three joints of the SCARA can be clearly depicted from the figure as well as their actuation blocks and the links of the robot. This model is later used in simulations to draw a square on the ground at a constant speed to test various controllers in Chapter VII.

3.6. Conclusions

Today most of the manufacturers utilize CAD models extensively to design and build their systems. Pressure on manufacturers for quick turn-around times encourages new methods for rapid prototyping techniques. Hence, integration of CAD models into other professional software packages facilitates the design process.

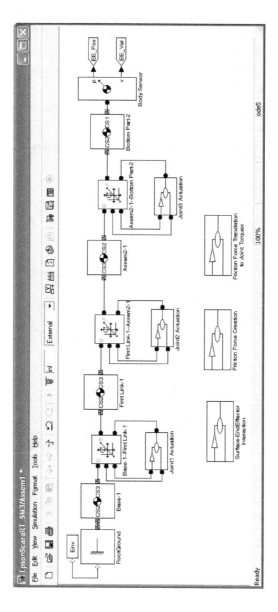

Figure 3.25. Simulink blocks of the Epson E2H853C

To streamline the design process for robot design, this work addressed the use of SolidWorks© and Matlab© environments. Such a virtual rapid prototyping approach allows design engineers to simulate the robot motions, check its workspace, design a suitable controller, test its performance, and modify the design at any stage of the design process before even building a prototype. Smooth and accurate information flow between various modules is seen as a crucial step to save time as well as prevent errors.

The virtual rapid prototyping approach is seen especially beneficial for robots, which enjoy a high number of actuated axes; therefore, a high number of actuators need to be controlled in unison. The system model development process is demonstrated in terms of RP, SCARA, and 6R serial manipulators although the methodology is valid for all parallel and serial mechanisms. This methodology to create simulation models is utilized for all the simulation studies in this work.

CHAPTER IV

DEVELOPMENT OF VIRTUAL HAPTIC ENVIRONMENTS

FOR REAL-TIME EXPERIMENTS

4.1. Introduction

The concept of Virtual Rapid Robot Prototyping (VRRP) was earlier introduced in Chapter III to model teleoperation systems for simulation studies. In this chapter, actual master systems (manual controllers) are integrated to the system using Matlab© tools to build simulations. This is necessary to run real-time experiments for the developed teleoperation controllers.

The development of virtual haptic environment is explained here by creating a Virtual Reality (VR) model for a robotics research laboratory along with all the existing robotic devices available in the lab. This virtual environment is then used to control these robotic devices (slave systems) through existing master systems such as force-reflecting joysticks and steering wheels. The VR model is built as a haptic environment to send force information to the operator (master system) as the remotely controlled device approaches a virtual obstacle.

The developed environment can also be used for training purposes. Hence, the operator can control the actual robot after having the necessary experience and training with the VR system. Simulators are widely used to train aircraft pilots [68], military [69] and medical personnel [70], and in entertainment industry. These simulators are mostly composed of the replica of the actual system with a VR screen to create the most realistic experience for the operators. The effectiveness of the VR training has captured the interest of many researchers in medical field [71], military [72] and aircraft pilot training [73]. Although researchers still discuss the effectiveness of VR simulators, other industries such as entertainment and automotive [74] started to use these simulators for entertainment and training purposes.

The architecture of the simulators is very similar to teleoperation system configuration. There is a controller device that the human operates with and a device that is driven by the demands from this controller. In teleoperation, the controller device is referred as the master system while the controlled device is called the slave system. The next section reviews some of the simulators that are used in industry. These simulators can also be considered as examples to basic teleoperation systems.

In the third section, robotic systems in the Robotics & Automation Laboratory at FIU are introduced as either the controller or slave devices. Later, integration of these devices in a virtual environment is explained. Finally, sample virtual haptic models of a Civil Engineering Laboratory, and the Robotics & Automation Laboratory at FIU are presented.

4.2. Simulator Background

The use of simulators is not limited to the application areas presented in the previous section. A group of researchers have incorporated a biomedical engineering application in simulator design to develop a flight simulator using the electromyographic (EMG) signals [75]. The simulator is shown in Figure 4.1. In this figure, the simulator is depicted on the left picture, the computer interface window to control the simulator is in the middle, and the picture on the right shows the connection of the EMG signals to the computer. The simulator is composed of a cockpit that is placed on a Stewart platform as commonly found in today's commercial flight simulators.

Miner and Stansfield utilized voice recognition in their version of a VR simulation system for robot control [76]. The demands to drive the robot are spoken by the operator. These demands are processed into commands and fed to the robot. As the robot carries out a task, it also sends out audio effects to interact with the operator.

Figure 4.1. Flight simulation using EMG signals [75]

A training simulator for VR arthroscopy is developed by German researchers [77]. Traditionally, the experience to perform an arthroscopy was gained by observing an experienced surgeon in operation. These researchers offer an effective alternative for training and establishing arthroscopic techniques in their work. Two pictures from the arthroscopy training simulator are given in Figure 4.2.

In the past, many simulators are developed for the automotive industry that are used for training and testing purposes. One such work has developed a PC-based driving simulator using Matlab© [78]. The driving simulator cockpit in this work is shown in Figure 4.3 with a trainee controlling the system via a steering wheel. This system is composed of the cockpit to send commands to the data acquisition card in the PC and the VR screen to display visual feedback to the driver. The system architecture with these components is shown in Figure 4.4.

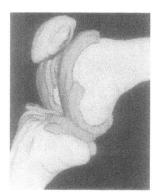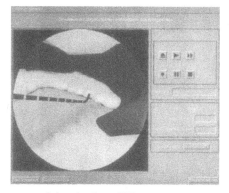

Figure 4.2. Arthroscopy training simulator [77]

Researchers from Korea have also developed a driving simulator which they integrated into a Stewart platform under the cockpit to imitate real world navigation [79]. They also used audio feedback as part of the VR screen.

Figure 4.3. Driving simulator cockpit [78]

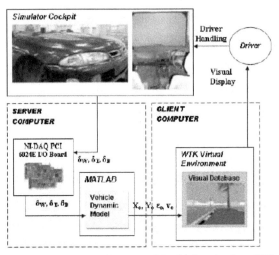

Figure 4.4. System architecture of the driving simulator [78]

Although some of the systems found in the literature are simpler than others, the ultimate task is to provide training for the operator before using the actual system as realistically as possible. The main difference between the two driving simulators mentioned above is that one of the simulators uses a Stewart platform which provides a more realistic experience to the trainee than the other. Therefore, addition of more feedback information than visual, such as audio, force, etc., enhances this experience.

4.3. Description of Robotic Systems in the Laboratory

There are two sets of robotic systems in the Robotics & Automation Laboratory. The first set includes robotic controller devices that are used with the PC as part of the human-computer interface. These devices include a gimbal-based force reflecting joystick, a force-reflecting steering wheel, a force feedback joystick and a Phantom Omni Device from SensAble Technologies.

Robots that are driven by the commands received from the human-computer interfaces form the second set of robotic systems in the laboratory. This type of configuration is very similar with teleoperation applications. Therefore, the human-computer interfaces can be titled as master systems or controller devices, and the driven devices as slave systems. In this work, the slave systems are listed as Motoman UPJ industrial arm, a holonomic mobile platform, and two mobile platforms; WiRobot DRK8000 and X80, by Dr. Robot, Inc.

51

4.3.1 Controller Devices

Master systems (controller devices) form a part of the human-computer interface. The operator sends the demands through the master system and the computer receives these demands though an interface. In this work, this interface is developed in Matlab$^{©}$ and C++. As the actual or virtual slave interacts with the environment, the created forces are then sent back to the actuators of the master in order to create a haptic environment. Therefore, all the master systems used in this work are selected not only to issue demands as the human operates on them but also to resist operator's motions as ordered by the slave side's interaction forces.

The first slave system presented here is a two degree-of-freedom gimbal-based force-reflecting joystick that was designed and manufactured in the Robotics and Automation Laboratory [80]. Each degree-of-freedom is bedded in between two servomotors. Each degree-of-freedom is uncoupled since the joystick has a gimbal-based design. The interface of this joystick is developed in C++ and later formatted into s-function blocks to be used in Matlab$^{©}$ Simulink within the Virtual Laboratory Simulator. The configuration of the joystick is shown in Figure 4.5.

Figure 4.5. Two-DOF master joystick

The second master system is a force-reflecting steering wheel. This system is a commercial gaming steering wheel from Genius [81] with acceleration and brake pedals and force feedback option. It also has twelve pushbuttons that enhance the interaction of the operator with the VR screen. It is primarily intended to operate mobile platforms. The steering wheel is shown in Figure 4.6.

Figure 4.6. Genius steering wheel [81]

The third master system is a commercially available force feedback joystick from Logitech (Wingman Force 3D) [82]. This system is capable of observing the motion about three axes and also reflecting the forces in two axes. It also has seven pushbuttons that enhance the interaction of the operator with the VR screen. The joystick is shown in Figure 4.7.

Figure 4.7. Logitech force feedback joystick [82]

53

Handshake VR Inc. offers a generic interface for joysticks in its software package, proSENSE[TM]. This interface is used to integrate the steering wheel to the Simulink environment. The interface provides the changes in the position of the wheel, pedals, and the pushbuttons of the Genius steering wheel or changes in position along the three axes of the Logitech joystick. It is also capable of sending force feedback information to the steering wheel or the joystick. The force feedback information is used to drive the actuators of the steering wheel or the joystick. The interface block and block parameters are shown in Figure 4.8. In the parameter window, the joystick identification number can be placed if more than one joystick is used. The designer can also specify the number of axes and pushbuttons to be monitored. Also, the number of output axes can be chosen in order to send force feedback information.

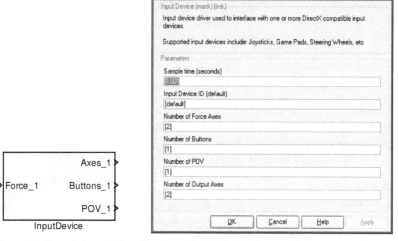

Figure 4.8. Generic joystick interface from Handshake VR Inc. and its parameter window

The last master system is a six degree-of-freedom Phantom Omni Device from SensAble Technologies as shown in Figure 4.9. Joint position readings are obtained through the encoders of each joint and the slave is driven with these inputs. However, only the forces can be reflected back to the operator with this device. Therefore, only the interaction forces created in the x, y and z directions are felt by the operator.

54

Figure 4.9. Phantom Omni Device from SensAble Technologies [83]

Handshake VR Inc. also provides a Matlab[©] interface for the Phantom Omni Device. The interface can get the position information from all joints of the device and it can also provide the position and orientation of the end-effector. This interface can also reflect the forces created as a result of the interaction.

4.3.2 Slave Robots

The robots that execute the tasks demanded by the operators are called slave robots. One of these robots is a holonomic mobile platform designed and manufactured in Robotics and Automation Laboratory [84]. The platform has four omni-directional wheels. These wheels enable the platform to move in any direction at any orientation. Therefore, the motion along any axis is uncoupled. The platform also has range sensors to sense the obstacles on its path. This information is then transformed into force feedback to drive the motors of the controller. Consequently, the operator feels that the platform is at the vicinity of an obstacle as a result of the resistance in the motion of the controller. Figure 4.10 shows the holonomic platform at its design stage. The platform has the capability to communicate with the host computer through a Bluetooth connection.

55

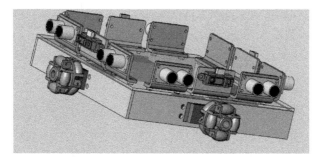

Figure 4.10. Holonomic mobile platform

The next set of mobile robots integrated into the virtual laboratory is the commercially available systems from Dr. Robot, Inc. The first system in this set is the X80 as shown in Figure 4.11. It has two 12V DC motors that supply 300 oz.-inches of torque to its 18 cm (7 in.) wheels, which yield a top speed of 1 m/s (3.3 ft/s). There are two high-resolution (1200 count per wheel cycle) quadrature encoders mounted on each wheel to provide high-precision measurement and control of wheel movement. The platform weighs 3.5 kg (7.7 lb.), and can carry an additional payload of 10 kg (22 lb.) as provided in its specification sheet [85].

Figure 4.11. X80 mobile platform from Dr. Robot, Inc. [85]

X80 communicates with the host computer through the WiFi (802.11b) wireless communication protocol. It has collision detection sensors that include three sonar range sensors and seven infrared range sensors. The information induced from these sensors are first transformed into force information and then sent back to the actuators of the controllers. It has also a camera that is operated with two servomotors, which can be used for image processing tasks and a microphone that can be utilized in applications that require voice recognition.

The second system is the WiRobot DRK8000 as illustrated in Figure 4.12. The system runs on a wheel-based platform with rotary sensors mounted on each wheel to precisely measure and control wheel motions. The eyes (webcams) can pan and tilt and the neck can pan and tilt as well [85].

Figure 4.12. WiRobot DRK8000 mobile platform from Dr. Robot, Inc. [85]

Communication with the host computer is accomplished through Bluetooth connection. The operator is able to send commands at a rate of 10-50 Hz. There are six ultrasonic range sensors and one infrared range sensor placed around the platform to detect obstacles. This sensory information is also sent back to the host computer as force information to be fed into the actuators of the controllers. This platform also has a camera and microphone, which enables it to be used for tasks involving image processing and voice recognition.

Motoman UPJ industrial arm, as shown in Figure 4.13, is the final robotic device used as a slave system in this work. It is different from the previous systems as it has a predefined limited workspace. It is a six-revolute (6R) joint manipulator that can work in three-dimensional space. It is commonly used in small part handling, assembly, automation, inspection, testing, education, and research applications.

Figure 4.13. Motoman UPJ industrial arm [86]

4.4. Haptic Virtual World Creation

Simulators with VR screen have to have two main components of the actual system. The actual systems for this case are the slave robots introduced in the previous section. The first component is the visual representation of the system that is created in a three-dimensional medium. Three-dimensional Computer-Aided-Design (3D CAD) tools are the most commonly used tools to accomplish this.

The second component is the modeling of the dynamics of the moving parts. Most of the time, the forward dynamics of the systems that are in motion are developed at another programming platform than 3D CAD such as in C++ or Matlab$^©$ Simulink. The Virtual Rapid Robot Prototyping idea presented in the previous chapter provides a more convenient option. A translator is used to extract the physical properties from the CAD model into the Matlab$^©$ Simulink environment. These physical properties include the mass, inertia, joint and center of gravity information, link lengths, and other geometric parameters of the system.

The visual representation created in a 3D CAD environment is also translated into the VR format and used with the Matlab$^©$ Virtual Reality Toolbox. This procedure was earlier explained in Chapter III. Although dynamics and visual representation of the system is

58

extracted from the same 3D CAD drawing into Matlab$^©$, these two have to be linked. As the dynamics simulation runs, the results of the simulation (joint positions, end-effector position) are fed into the Virtual Reality block. Therefore, visual representation of the system is actuated with the results obtained from the dynamic simulation.

There are two ways to create the haptic effect in this setup. The first and the tedious way is to construct the obstacles inside the dynamic simulation using the logic relations. This method is feasible for simple tasks that involve very limited amount of obstacles. The second method is to extract the obstacle information from the VR model. This model contains all the necessary information on the surfaces of the obstacles. Handshake VR Inc. has a commercially available toolbox, proSENSE^{TM}, which extracts the obstacle information off the VR model and also creates force for point type of contact. This way, the force feedback information can be obtained even for very large scale simulations without developing extensive logic operations manually.

Up to this point, creation of a haptic simulation in Matlab$^©$ is explained. The goal of this study is to create a VR model of a laboratory with haptic effects, for which the operator can interact with through a human-computer interface. In the previous section, the controllers to be used in this work have been introduced. The next step is to integrate these controllers with the haptic simulation environment.

The interface for the commercially available joysticks (Genius Steering Wheel) and the Phantom Omni Device is provided by Handshake VR Inc. as a Matlab$^©$ Simulink block. This interface communicates with the external device to receive position information from the encoders and send back force information to the actuators of the device.

An interface had to be developed for the gimbal-based joystick developed in the Robotics and Automation Laboratory. This interface communicates with the motion controller card that drives the servomotors of the joystick. The source code for this interface is given in [80]. Although the interface is written in C++, it is then formatted into a Matlab$^©$ Simulink block to have the same functions described in the previous paragraph.

As the interface is integrated to interact with the dynamic simulation, Real-Time Window Target option of Matlab$^©$ is utilized to synchronize the procedure with the real-time clock. Therefore, the created system is no longer a simulation but a VR simulator.

4.5. Sample Virtual Haptic Laboratories

Teleoperation is one of the research areas of the Robotics and Automation Laboratory at FIU. Various experiments are conducted by using an actual joystick and one of the slave robots. The slave robots are used as their virtual representations in these experiments. This type of architecture is very similar with the concept of developing a virtual haptic laboratory. As a matter of fact the only difference is the scale of the experiments. In teleoperation experiments, only one slave was used for a single task. By contrast, in the virtual haptic laboratory, a user will be able to operate any of the slave robots through any of the controllers.

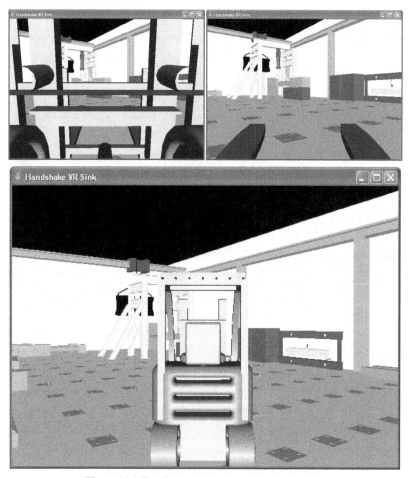

Figure 4.14. The three camera views from the forklift

The first trial for a virtual haptic laboratory is constructed for a Civil Engineering laboratory at FIU. The task was to drive a forklift inside the laboratory while virtual experiments are conducted.

The forklift is controlled though the Genius steering wheel. The steering wheel controls the rotation of the forklift while the pedals are used to accelerate and decelerate the vehicle. Two pushbuttons are addressed to shift the gear up or down. Therefore, the vehicle is able to use multiple speed levels as well as a neutral position for the shift and a reverse gear. There are also three virtual cameras placed in the vehicle to view the virtual environment. The first camera is placed in the operator's seat, the second one is placed at the tip of the forks and the last one is placed at the top back part of the vehicle to have a general view of the laboratory as well as the vehicle itself. Three views from these cameras are shown in Figure 4.14.

The virtual experiments that involve applying pressure to the test specimens are controlled through the other pushbuttons of the steering wheel. Another commercially available joystick from Logitech is used to control the crane that carries a work specimen inside the lab. The crane has the architecture of a three degree-of-freedom Cartesian robot therefore all three axes of the joystick are used to operate this crane. Three pushbuttons are also activated to apply and release the breaks for each degree-of-freedom.

After developing a successful but rather simple virtual laboratory simulator, the next step is to configure the Robotics and Automation Laboratory in the Virtual World. 3D CAD drawing of the laboratory is constructed in SolidWorks©. Initially, the WiRobot DRK8000 is modeled to navigate around the laboratory. In order to achieve this, the robot parameters are translated into Matlab©. Motoman UPJ arm is also placed in the drawing for future experiments.

Genius steering wheel is selected to control the mobile platform. The camera is fixed to the location of the webcam of the actual robot. Therefore, the operator will experience a more realistic training period with the simulator. The head of the robot that carries the webcam moves as it receives commands from the pushbuttons of the steering wheel. As a result, the view changes as the head moves in different directions. This is in fact what an operator will experience while using the actual robot. The only visual feedback is through the webcam installed at the head of the robot. This will train the operator not to lose the orientation while navigating around the laboratory as the head is in motion. Figure 4.15 shows a few of screenshots of the VR screen as the robot navigates around the laboratory.

A snapshot view of the robot is shown in Figure 4.16. The haptic effects are created using the Handshake VR Inc. software and fed into the actuators of the steering wheel as the robot approaches to an obstacle. This effect is accurate for the navigation of the actual robot as well. The range sensors of the robot create haptic effects and send this information to the steering wheel in the same fashion of the simulator.

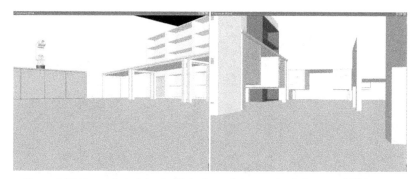

Figure 4.15. Screenshots from the Robotics and Automation Laboratory VR Screen

Figure 4.16. Screenshot of the WiRobot DRK8000 inside the Robotics and Automation Laboratory VR Screen

Figure 4.17 shows the operator driving the DRK8000 mobile platform inside the virtual representation of the Robotics and Automation Laboratory. The Motoman UPJ arm can also be depicted from the VR screen in this figure.

Figure 4.17. Operator driving WiRobot DRK8000 in VR screen of the Robotics and Automation Laboratory

4.6. Conclusions

The concept introduced in the previous chapter can be applied to model any environment ranging from undersea to space and surgery rooms. These models can then be used for simulation studies. Also, any PC-based controller device can also be integrated to the model by using the tools described in this chapter. Therefore, using the VRRP and the method to develop virtual haptic environments, a wide range of experimental setups can be created. These setups find use in simulators to train personnel and also in teleoperation experiments.

One of the goals of the work presented in this chapter is to configure real-time experimental teleoperation setups quickly and accurately. This method is used to build all the real-time experimental setups described later in this book. It has also provided a useful platform to test various controllers, and compare the performance of different controller devices. The equipment to configure these experimental setups is introduced in the following chapter.

As mentioned above, another goal is to create a general purpose simulator tool to rapidly train the human teleoperators before using the robotic devices that may be positioned in any environment. The operator will thus gain experience training in the virtual environment without damaging the equipment or harming personnel. In fact, today most of the system developers in military, aircraft and medical industries use simulators developed for specific systems and environments for training purposes.

The simulator tool developed in this work offers a capability to create any environment and then integrate different controller devices (master systems) and remote robots (slave systems) for simulation as well as real-time operations.

CHAPTER V

FAULT TOLERANT DESIGN

5.1. Introduction

Teleoperation systems are often employed in critical missions that take place in hazardous or unreachable sites for humans. While the operator works with the master system in a safe environment, the slave system accomplishes the task in a remote, usually hazardous, environment. Robots are generally considered as complex electromechanical devices. One of the electromechanical components can always fail during a critical task such as a space mission, telesurgery, or bomb disposal. These are few critical tasks where the slave robot is still required to complete its mission even when a component fails. Fault-tolerant design offers a solution for these types of applications by enabling the robot to continue the mission even if its actuators, sensors or link mechanisms fail.

Teleoperation design engineers need to evaluate the task correctly to decide on the necessity of fault tolerance. If the task takes place at a site unreachable or hazardous to humans, and the mission is required to be continuous even when one part of the robot fails, design engineers need to modify their designs to have fault tolerance capabilities.

Fault tolerance can be achieved at different levels as explained in Chapter II. If the communications line of the system has inconsistencies, then this type of failure can be overcome by fault-tolerant controller employment. The controller architectures of this type are presented in the next chapter. When the failures are expected to occur in master and slave systems, it is clear that the systems need to be designed with fault tolerance capabilities.

The system can be designed to have sensor/computer, joint or mechanism level fault tolerance features or a combination of them. Fault tolerance can also be achieved by having a second identical system to take over the mission if the first one completely fails. Another way of achieving fault tolerance in a teleoperation system is to employ manipulators that are redundant for the defined task. For instance, regular six-DOF robotic arms can be used as fault-tolerant robots in tasks where the orientation of the end-effector is less important, or tasks take place in a plane. Therefore, if a link fails during the mission, the teleoperation can remain continuous by degrading the orientation tracking performance with the remaining five links.

This chapter presents all the master and slave systems used in real-time experiments. These systems are also configured to have fault tolerance by different means. One of the

65

master and two of the slave systems have fault tolerance in certain levels by design. In this case, the master system is the gimbal-based, two-DOF joystick, and the slave systems are a holonomic mobile platform and the replica of the gimbal-based, two-DOF joystick. The rest of the master and slave systems are commercial robotic systems with different configurations and DOF. These manipulators are not fault tolerant at the sensor, computer or joint level. Hence, if the tasks described for them require fewer DOF, then they are considered redundant manipulators for such tasks with mechanism level fault tolerance.

In this chapter, the master and slave systems are described in the following order: Two-DOF Master Joystick, Phantom Omni Device, Two-DOF Slave Joystick, Holonomic Platform, Epson SCARA arm, Fanuc arm, and Motoman arm. As the slave systems are introduced, their virtual models are also presented as they are utilized in teleoperation experiments.

5.2. Gimbal-Based Master Joystick

The first master system presented here is a two-DOF, gimbal-based force-reflecting joystick that was designed and manufactured in the Robotics and Automation Laboratory. Both DOF are composed of revolute joints and the gimbal-based design of the joints enables uncoupled motion between the two DOF. Each joint is designed to be bedded in between two servomotors. Hence, joint level fault tolerance is achieved through this architecture. If one of the servomotors fails during the manipulation, the other servomotor completes the task. The configuration of the joystick with its x and y rotation axes is also shown in Figure 5.1.

Each servomotor has an encoder connected to the rear end of its shaft. The demands of the human operator are sensed through these encoders. The measured change in the position of the joystick is then transmitted to the slave as manipulation demands. The forces extracted from the slave system are forwarded as torque commands to the servomotors to achieve telepresence. The specifications of the servomotors used are given in Table 5.1 [87]. The dimensions of the servomotor are presented in Figure 5.2.

The motion controller, DMC 18x2, from Galil Motion Control, Inc. [88] is used for the control of the servomotors through the computer. The shortcoming of this controller for our test environment is that it does not have an interface for Matlab©. In other words, Matlab© does not support this motion control card for the real-time applications. Therefore, an interface was created [80]. This interface sends out torque information to drive the servomotors of the joystick and receives encoder readings to gather the position information of the joystick.

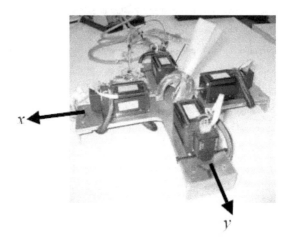

Figure 5.1. Two-DOF master joystick with its rotation axes

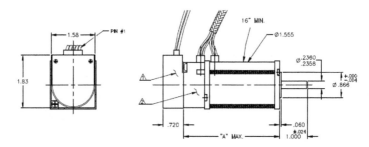

Figure 5.2. Pittman ELCOM 4441S010 servomotor dimensions [87]

This interface is written in C++ to be used as an s-function block in the Matlab$^©$ model. This block is shown in Figure 5.3. As a result of this, necessary information to run the teleoperation system is sent back and forth between the model and the actual joystick by this interface. Similar studies have been carried out for serially-connected microcontrollers [89]. For this study, the PCI card from Galil is targeted to communicate with the servomotors through the Galil amplifier system.

In Figure 5.3, the interface provides torque information for each axis of the joystick which is sent from the blocks called "Force Feedback X" and "Force Feedback Y." The servomotors are driven using this information. Another demand is sent to set the encoders of the joystick to pull it to null position. This is required to start the test from the null position since the encoders are not absolute. The other two inputs used in the joystick are to disable the servomotors and terminate the communication between the real-time system and the simulation. These are required for security reasons and to prevent the servomotors from overheating under continuous use.

Table 5.1. Specifications of the Pittman ELCOM 4441S010 servomotor

Parameter	Sym	Unit	ELCOM
Continuous Torque Max	T_C	oz·in (N·m)	12.00 (.084)
Peak Torque-Stall	T_{PK}	oz·in (N·m)	71 (0.5)
Friction Torque	T_F	oz·in (N·m)	0.15 (1.1×10^{-3})
No Load Speed	S_{NL}	rpm (rad/s)	5780 (605)
Rotor Inertia	J_M	oz·in·s^2 (kg·m^2)	6.4×10^{-4} (4.5×10^{-6})
Electrical Time Const.	τ_E	ms	0.18
Mechanical Time Const.	τ_M	ms	5.5
Viscous Damp. - Infinite Source Imp.	D	oz·in/krpm (N·m/(rad/s))	0.038 (2.6×10^{-6})
Damping Const— Zero Source Imp.	K_D	oz·in/krpm (N·m/(rad/s))	12.3 (8.3×10^{-4})
Max Winding Temp	θ_{MAX}	°F (°C)	266 (130)
Thermal Impedance	R_{TH}	°F/watt °C/watt	44 (6.7)
Thermal Time Const.	τ_{TH}	min.	22.8
Motor Weight	W_M	oz (Mass) (g)	17 (482)
Motor Constant	K_M	oz·in/√W (N·m/√W)	4.07 (.0287)
Motor Length	L_1	in max. (mm max.)	4.375 (111.1)

Only joystick positions in x and y axes are received through the interface into the simulation environment as "Pos_FB_1" and "Pos_FB_2." Joystick position information is then scaled with the scale factors to be fed to the slave. The source code of the interface is provided in Appendix A.

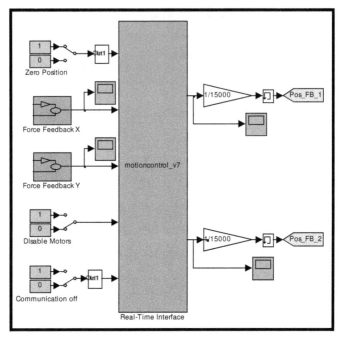

Figure 5.3. Interface window for the 2-DOF joystick

5.3. Phantom Omni Device

The second master system is the Phantom Omni Device from SensAble Technologies, Inc. as shown in Figure 5.4. It is a PC-based six-revolute-joint (6R) arm with an IEEE 1394 (FireWire) interface for the PC connection. All the joints have encoders attached to their shafts and joint position readings are obtained through the encoders. However, only the forces can be reflected to the operator with this device. Therefore, only the interaction forces created in the x, y and z directions are felt by the operator and the torque information about these axes are disregarded. The specifications of the device are listed in Table 5.2 [83].

Phantom Omni Device is used to control the Motoman UPJ industrial arm in the laboratory. It can also be used to control the mobile platforms in the laboratory with the appropriate mapping of the motions.

For this device, Handshake VR, Inc. provides the Matlab© Simulink interface. The interface block and the block parameters are presented in Figure 5.5. From the parameter window, the designer can select to read the joint positions as well as the position and orientation of the end-effector in Cartesian space. This information is then utilized by wiring the necessary blocks to the out-ports that are illustrated in the "OmniDevice" block (Figure 5.5). The user can also gather the on/off information for both pushbuttons of the device from the out-port "Buttons." The only in-port channel is reserved for the force feedback information to be fed to the actuators of Phantom.

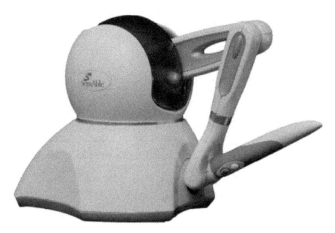

Figure 5.4. Phantom Omni Device in its initial configuration [83]

Table 5.2. Specifications of the Phantom Omni Device

Force feedback workspace	~6.4 W x 4.8 H x 2.8 D in > 160 W x 120 H x 70 D mm
Footprint (Physical area device base occupies on desk)	6 5/8 W x 8 D in ~168 W x 203 D mm
Weight (device only)	3 lbs 15 oz
Range of motion	Hand movement pivoting at wrist
Nominal position resolution	> 450 dpi ~ 0.055 mm
Maximum exertable force at nominal (orthogonal arms) position	0.75 lbf (3.3 N)
Continuous exertable force (24 hr)	> 0.2 lbf (0.88 N)
Stiffness	X axis > 7.3 lbs/in (1.26 N/mm) Y axis > 13.4 lbs/in (2.31 N/mm) Z axis > 5.9 lbs/in (1.02 N/mm)
Force feedback	x, y, z
Position sensing	x, y, z (digital encoders)
Stylus gimbal	Pitch, roll, yaw (± 5% linearity potentiometers)
Interface	IEEE-1394 FireWire® port

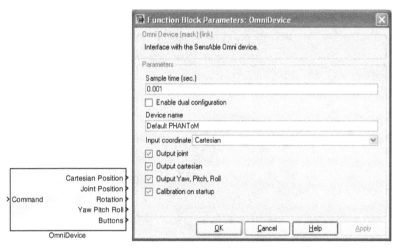

Figure 5.5. Interface block of the Phantom Omni Device and its block parameters

5.4. Gimbal-Based Slave Joystick

The first slave is constructed as the replica of the two-DOF gimbal-based joystick in virtual environment. The virtual representation of this slave and its forward kinematic and dynamic models are created using the VRRP method. The virtual representations of the joystick are shown in Figure 5.6.

This slave is used in identical master-slave experiments along with the actual gimbal-based joystick. As a result of this, mapping is not required for the motion demands received from the master. Instead, it receives a demand for each joint directly and independently. The Simmechanics model of the system is shown in Figure 5.7.

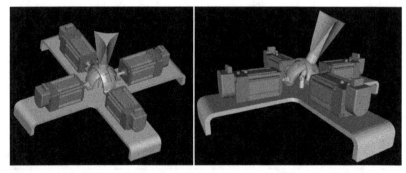

Figure 5.6. Virtual representation of the identical slave joystick

71

In this model, the joystick base is fixed to the ground which disables its motion relative to the ground. "Joystick_Base" is composed of the inertia and mass information of the stators of the servomotors and the aluminum platform that is used to mount these servomotors. The gimbals of the joystick are connected to the base with the revolute joints that rotate about the x and y axes. These joints are driven by the actuation blocks placed under them. The gimbals are then attached to the stick of the joystick to simplify the model. "Gimbal X" and "Gimbal Y" bodies are composed of the gimbals and the rotors of the servomotors. As it can be depicted from the block diagram, the two DOF are working as separate mechanisms that do not interact. This enables the joystick to have uncoupled motions about both axes.

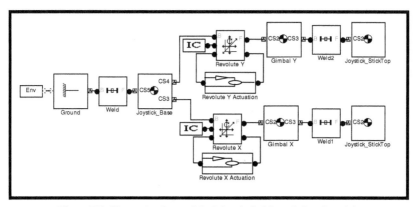

Figure 5.7. Simmechanics model of the gimbal-based slave joystick

5.5. Holonomic Mobile Platform

The second slave system described in this chapter is the holonomic mobile platform which has an unlimited workspace. Omni-directional wheels are used to develop a holonomic type mobile robot. These wheels enable the robot to move in any direction at any orientation. There is no need to change the orientation of the platform while traveling in an arbitrary trajectory. The direction of the linear velocity is independent from the orientation of the vehicle. Ultimately, this system may be considered as a three degree-of-freedom planar robot.

The holonomic mobile platform has fault-tolerant architecture. It has fault tolerance in link level and also has Triple Modular Redundancy (TMR) in its sensor configuration. Link level redundancy is provided using four independently actuated wheels for a three degree-of-freedom motion. The virtual representation and the actual picture (without TMR in its sensor

72

configuration) of the platform is shown in Figure 5.8. The design stages of this robot and the fault tolerance features are described in detail in Appendix B.

Figure 5.8. Virtual representation (l) and the actual picture (r)
of the holonomic mobile platform

The platform is intended to be used with the two-DOF gimbal-based joystick in teleoperation experiments. As a result of this, only two DOF of the platform is used. The model is developed for only the motions along two axes provided that there is no slippage; hence, the orientation of the platform does not change during the telemanipulation. Figure 5.9 shows the holonomic mobile platform model.

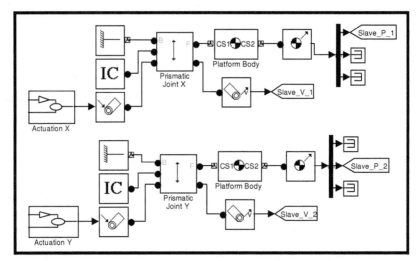

Figure 5.9. Simmechanics model of the holonomic mobile platform

The model can be observed as two prismatic joints working independently. This is the case for holonomic motion for mobile platforms. The motion along one direction is not affected by the motion along the other. The Simulink "Platform Body" blocks are the same for both joints since it represents the mass and inertia of the platform and both joints actuate the same body. This platform receives position change information from the gimbal-based joystick and issues them as velocity changes in its "Actuation X" and "Actuation Y" blocks.

5.6. Epson SCARA Manipulator

This system is a four-DOF SCARA manipulator by Epson (model E2H853C). It has a limited workspace and requires mapping between the joystick motion and the slave motion. The specifications of the manipulator are given in Table 5.3 [90]. These specifications are used to model the robot.

Simulation model of this manipulator is also used in the real-time experiments. It is constructed as a virtual slave manipulator. The link parameters of the manipulator required for modeling purposes are given in Table 5.4. These parameters are also used in inverse kinematics solution of the manipulator. The motions of this manipulator in Cartesian space are mapped with the motions of the master. Later, using the inverse kinematics solution, the demands received in Cartesian space are translated into joint demands. The joints are then driven with these demands.

The VRRP method is used to construct the robot in virtual environment. First, the manipulator is constructed in a computer-aided-design software environment. Then, the material, inertial and mechanism parameters are translated into the Matlab$^©$ environment. Figure 5.10 shows a visual representation of the manipulator along with the actual manipulator.

The friction between the end-effector and the surface is created using Coulomb friction and stiction model. The coefficient of friction used in the experiment is of lead which is specified as 0.43 and stiction is regulated at 1.00. The model shown in Figure 5.11 includes end-effector surface interaction model and surface friction models. This required to model the interaction of the end-effector with the surface accurately. The friction forces that are created as explained previously are later fed in the joints as disturbances with the help of the "Friction Force Translation to Joint Torques" block. These blocks are not created automatically when the model in SolidWorks is translated into Simulink blocks. The designer has to write the code for these subroutines, or in the Simulink case, the designer has to wire the necessary blocks together to create a subsystem like "Friction Force Creation" block.

74

Table 5.3. Specifications of Epson E2H853C Manipulator

Arm length	Arm #1+#2	500 mm + 350 mm
Weight (not include the weight of cables)		37 kg; 83lb
Driving method	All joints	AC servo motor
Max. operating speed	Joint #1+#2	5266 mm/s
	Joint #3	1100 mm/s
	Joint #4	1428 deg/s
Repeatability	Joint #1+#2	±0.025 mm
	Joint #3	±0.01 mm
	Joint #4	±0.03 deg
Max. motion range	Joint #1	±145 deg
	Joint #2	±147 deg
	Joint #3	290 mm
	Joint #4	±360 deg
Max. pulse range	Joint #1	-62578 to +267378
	Joint #2	±133803
	Joint #3	-85525 to 0
	Joint #4	±86016
Resolution	Joint #1	0.0008789 deg/pulse
	Joint #2	0.0010986 deg/pulse
	Joint #3	0.0033908 mm/pulse
	Joint #4	0.0041852 deg/pulse
Motor power consumption	Joint #1	400 W
	Joint #2	400 W
	Joint #3	400 W
	Joint #4	150 W
Payload	rated	2 kg
	max.	20 kg
Joint #4 allowable moment of inertia	rated	0.02 kg.m^2
	max.	0.45 kg.m^2
Joint #3 down force		200N

Table 5.4. Link Parameters of Epson E2H853C Manipulator

Joints	α_k (deg)	s_k (mm)	a_k (mm)	θ_k (deg)
1	0	0	500	θ_1
2	0	0	350	θ_2
3	0	s_3	0	0
4	0	0	0	θ_4

α_k : Twist angle between the axes of the joints k and $k+1$

s_k : Axial offset along the axis of the joint k

a_k : Common normal distance between the axes of the joint k and $k+1$

θ_k : Angular position of the kth link with respect to the $(k-1)^{th}$ link

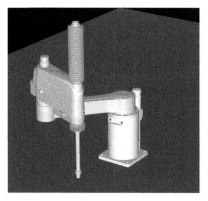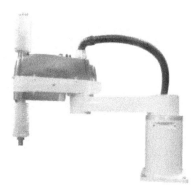

Figure 5.10. Virtual representation of the SCARA robot (l)
accompanied by the actual manipulator [90] (r)

As observed from the block diagram, only three joints of the SCARA are used in the
model. The first two joints are simple revolute joints that rotate about the z-axis. They are
actuated with the outputs from "Joint1 Actuation" and "Joint 2 Actuation" blocks. The third
joint is a cylindrical joint which is composed of a revolute joint and a prismatic joint.
Revolute joint is used to control the orientation of the end-effector, whereas prismatic joint is
responsible for the motion along the z-axis. The task designated for this manipulator does not
require tracking the orientation; therefore, only the prismatic joint of the cylindrical joint is
driven. The block named "Joint 3 Actuation" drives the prismatic joint of the SCARA.

5.7. Fanuc Industrial Robot Arm

This slave system is another serial industrial robot arm, Fanuc LR Mate 100iB. The
Fanuc LR Mate 100iB is a five-axis, electric servo-driven robot. It is capable of a wide
variety of tasks in a broad range of industrial and commercial applications including machine
tending and part transfer processes. Fanuc describes this manipulator to have high joint speed
that maximizes throughput, a capability of flipping over backwards for a larger work
envelope, and absolute encoder positioning that eliminates homing at power-up [91]. The link
and joint parameters of the manipulator are presented in Table 5.5 whereas Table 5.6
tabulates the manipulator specifications.

The teleoperation task is defined as tracing horizontal surfaces maintaining a point
contact. While tracking the contour, the end-effector is required to maintain its orientation
parallel to the normal of the surface. Therefore, for the designed task only four DOF of the

76

manipulator are used. The last (fifth) joint is kept at a fixed position throughout the tests. First three joints are used for positioning while the fourth joint is used to maintain the orientation of the end-effector.

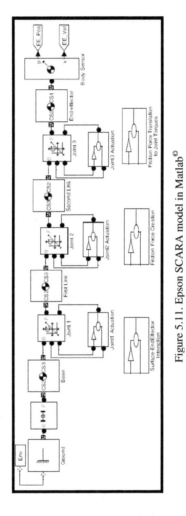

Figure 5.11. Epson SCARA model in Matlab©

During the manipulation, if any of the joints two, three or four fails, the orientation objective can be sacrificed but the position tracking can be maintained. This redundancy for the specific teleoperation task also promotes fault tolerance in the slave system. The actual manipulator and its virtual representation as used in the tests are shown in Figure 5.12.

Table 5.5. Link and joint parameters of the Fanuc LR Mate 100iB

Joints	α_k (deg)	s_k (mm)	a_k (mm)	θ_k (deg)
1	$-\pi/2$	0	151	θ_1
2	0	0	250	θ_2
3	0	0	200	θ_3
4	0	0	80	θ_4
5	$\pi/2$	0	0	θ_5

α_k : Twist angle between the axes of the joints k and $k+1$

s_k : Axial offset along the axis of the joint k

a_k : Common normal distance between the axes of the joint k and $k+1$

θ_k : Angular position of the kth link with respect to the $(k-1)^{th}$ link

Figure 5.12. Fanuc LR Mate 100iB [91] and its virtual representation

The serial arm described above is again integrated into the teleoperation system as a virtual representation of the original manipulator. That concept of VRRP is also used to construct the model for the Fanuc robot. The model of the manipulator is shown in Figure 5.13. Since the fifth joint is not used in the teleoperation experiments, only the first four joints were used in the model. The model also includes all the necessary custom created blocks, or subsystems, to build the interaction between the end-effector and the surface. These blocks were presented in the previous section. The "Inverse Kinematics Solution" block is used to map the motion demands received in Cartesian space into the joint space of the manipulator.

Table 5.6. Specifications of the Fanuc LR Mate 100iB

Controlled axes		5 axes
Max. load capacity at wrist		5kg
Motion range	J1	5.59rad (320deg)
	J2	3.23rad (185deg)
	J3	6.37rad (365deg)
	J4	4.19rad (240deg)
	J5	12.6rad (720deg)
	J6	-
Max. speed	J1	4.19rad/s (240deg/s)
	J2	4.71rad/s (270deg/s)
	J3	4.71rad/s (270deg/s)
	J4	5.76rad/s (330deg/s)
	J5	8.38rad/s (480deg/s)
	J6	-
Repeatability		+/-0.04mm
Mechanical unit mass		38kg
Application	Arc welding	x
	Spot welding	-
	Handling	x
	Sealing	x
	Assembling	x
	Others	Mold release spray Deburring
Remarks		Controller is R-J3iB Mate

5.8 Motoman Industrial Robot Arm

Motoman UPJ industrial arm is the final robotic device used as a slave system in this work. It is a six-DOF all-revolute-joint industrial robot arm and commonly used in small part handling and assembly, automation, inspection/testing, education, and research applications. The workspace and link lengths of the arm are presented in Figure 5.14.

The specifications of the arm are summarized in Table 5.7 [86]. At its current configuration, this industrial arm does not have any external sensors such as a range or force sensor. In order to create haptic effects, range sensing capability is added to the end-effector of the arm's virtual model.

79

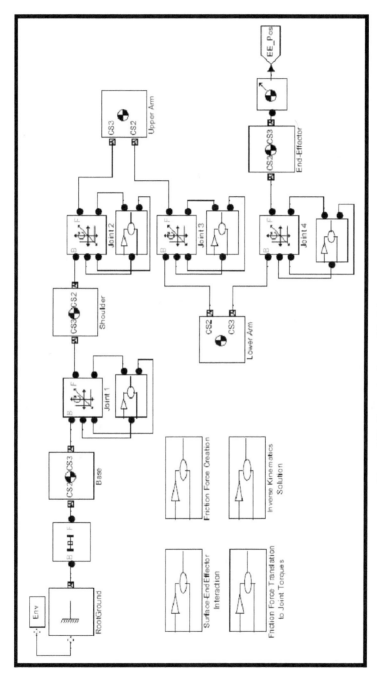

Figure 5.13. Simmechanics model of the Fanuc LR Mate 100iB industrial arm

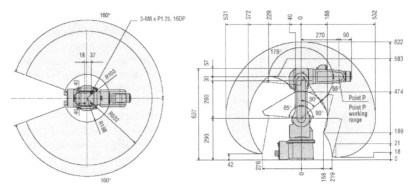

Figure 5.14. Workspace of the Motoman UPJ industrial arm [86]

This manipulator is also used as a virtual manipulator in the teleoperation experiments. The VRRP method is used to model the robot and create its virtual representation. The Motoman UPJ manipulator and its virtual representation are shown in Figure 5.15.

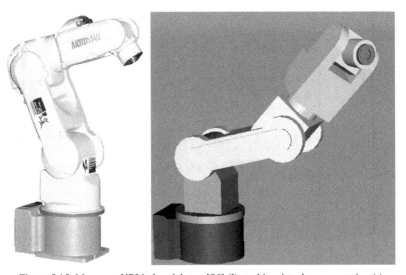

Figure 5.15. Motoman UPJ industrial arm [86] (l), and its virtual representation (r)

Table 5.7. Specifications of Motoman UPJ industrial arm

Controlled Axes		6
Payload		3 kg (6.6 lbs.)
Vertical Reach		804 mm (31.7")
Horizontal Reach		532 mm (20.9")
Repeatability		±0.03 mm (±0.001")
Maximum Motion Range	S-Axis (Turning/sweep)	±160°
	L-Axis (Lower Arm)	+90°/-85°
	U-Axis (Upper Arm)	+175°/-55°
	R-Axis (Wrist Roll)	±170°
	B-Axis (Bend/Pitch/Yaw)	±120°
	T-Axis (Wrist Twist)	±360°
Maximum Speed	S-Axis	200°/s
	L-Axis	150°/s
	U-Axis	190°/s
	R-Axis	300°/s
	B-Axis	300°/s
	T-Axis	420°/s
Approximate Mass		25 kg (55.1 lbs)
Brakes		All axes
Power Consumption		0.5 KVA
Allowable Moment	R-Axis	5.39 N.m (0.55 kgf.m)
	B-Axis	5.39 N.m (0.55 kgf.m)
	T-Axis	2.94 N.m (0.3 kgf.m)
Allowable Moment of Inertia	R-Axis	0.1 kg.m2
	B-Axis	0.1 kg.m2
	T-Axis	0.03 kg.m2

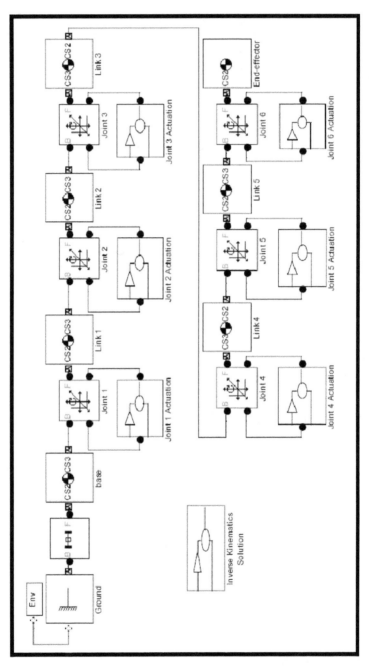

Figure 5.16. Simmechanics model of the Motoman UPJ

The model of this industrial arm is shown in Figure 5.16. All six joints are present in the model with their actuation blocks named "Joint X Actuation." This robot is controlled with the Phantom Omni Device in teleoperation experiments. As the Phantom Omni Device is also a 6R manipulator, this industrial arm can be used at its ultimate capacity, using all the DOF. The mapping between the two manipulators can be developed in joint or in Cartesian space. The "Inverse Kinematics Solution" block is used to map the demands to the joint space of the manipulator if they are received in Cartesian space. During the teleoperation, if one of the links of the robot fails, orientation tracking requirements can be sacrificed to continue the mission as good as possible.

5.9. Conclusions

In this chapter, all the robotic devices that are used in the real-time experiments are presented. The robotic devices with fault-tolerant designs as well as the ordinary industrial arms that do not have fault tolerance features are introduced. It is also explained that the industrial arms may be considered redundant relative to specific tasks that require less degrees of freedom than maximum available. Therefore, mechanism or link level fault tolerance is achieved for these manipulators.

The master systems are used as actual controller devices whereas slaves are used as virtual robots in the experiments. The Matlab$^©$ models of the slave manipulators are presented as well as specifications of the robots. These specifications are used to model the robots. The next chapter describes the teleoperation controllers used to integrate the master and slave systems introduced in this chapter.

CHAPTER VI

TELEOPERATION CONTROLLERS

6.1. Introduction

Bilateral teleoperation systems have a large variety of application areas. As they are used in diverse application areas, different feedbacks can be included in the systems as audio, visual, tactile and others. However, in this study only force feedback from the slave system to the master system is considered. This feedback actuates the master system's degrees of freedom to create a sense of force reflection so that the human operator feels the environment that the slave is working on. Therefore, control strategies that send motion information to the slave side and receive force information on the master side are considered.

Niemeyer in his dissertation suggests that passivity is not only sufficient but necessary for stability for teleoperation systems and presents a passive P.D. control for the basic teleoperation without time delays [22].

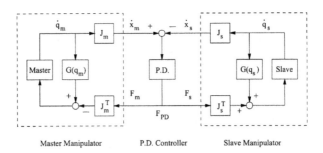

Figure 6.1. Master and slave systems that are connected with a P.D. controller mimicking a spring and damper [22]

Figure 6.1 shows the proposed P.D. controlled teleoperation design by Niemeyer [22]. J's are inertia matrices, $\dot{q}$'s are the velocities of the joints, $G(q)$'s are the gravitational compensators, F's are the forces and finally $\dot{x}$'s are the velocity components of the end-effector in the proposed block diagram.

However, in real-time teleoperation systems, there will always be some transport delay for the information transport between the master and the slave. Niemeyer studied instability in teleoperation systems due to time delays using the P.D. controller [22]. The next section describes modeling of time delays in control system design. Later, previously

85

proposed control algorithms that ensure the stability of teleoperation systems are presented. Following these control algorithm descriptions, the wave variable technique is introduced, as this method and its modifications are developed and used in this work. Finally, the proposed position/force control algorithms are introduced for use in the control schemes of the master and slave subsystems.

6.2. Modeling of Time Delays

In the systems that experience time delays, the input signal $x(t)$ is applied to a pure delay element, and the output of the signal will become a time-shifted version of the input. The signal is usually modeled not to have any distortion, and all frequencies already present in $x(t)$ pass without attenuation. The output signal is then described as $x(t-T)$ in which T is the time delay [92]. Assuming that $x(t)$ to be zero for $t<0$, $x(t-T)$ is zero for $t<0$,

$$L\{x(t-T)\} = \int_0^\infty e^{-sT} x(t-T)dt = \int_T^\infty e^{-sT} x(t-T)dt = e^{-sT} \int_0^\infty e^{-s\lambda} x(\lambda)d\lambda = e^{-sT} \overline{x}(s) \qquad (6.1)$$

where $\overline{x}(s) = L\{x(t)\}$ is the Laplace transform; hence,

$$L\{x(t-T)\} = e^{-sT} \overline{x}(s) \qquad (6.2)$$

This is the delay theorem of the Laplace transformation. Though it should be clearly stated that this result is of interest only for functions $x(t)$ which are zero for $t<0$.

Systems that are inherently by transcendental transfer functions are more difficult to handle in the analysis and design of control systems [93]. Many ways of approximating e^{-sT} by a rational function can be listed. One way of approximation is to use the Maclaurin series;

$$e^{-sT} \cong 1 - sT + \frac{s^2 T^2}{2} \qquad (6.3)$$

or

$$e^{-sT} \cong \frac{1}{1 + sT + s^2 T^2 / 2} \qquad (6.4)$$

in which just three terms of the series are used. The approximations are not valid when the magnitude of time delay, T, is large.

Padé approximation can also be used for a better approximation [94]. The two-term approximation is presented below:

$$e^{-sT} \cong \frac{1 - sT/2}{1 + sT/2} \tag{6.5}$$

In this work, Padé approximation is used in time delay blocks of Matlab$^©$ based simulations and experiments.

6.3. Time-Delayed Teleoperation Controllers

Munir listed the controllers for the time-delayed systems as Smith predictors, observer based design, and sliding mode controller [24]. Other than the listed controllers, Cho and Park [95] worked with impedance controllers for bilateral teleoperation.

6.3.1 Smith Predictors

Smith proposed a control scheme that allowed a high loop gain to improve accuracy. A feedback control system that experiences time delays under the control of Smith predictor design is shown in Figure 6.2.

$C(s)$ is the proportional controller and $P(s)$ is a first order filter with transport delay,

$$P(s) = \frac{e^{-sT}}{\varpi + 1} \tag{6.6}$$

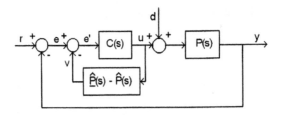

Figure 6.2. Feedback control system with the Smith predictor [24]

Smith proposed a minor correction loop around the controller as shown in Figure 6.2. As a result of this, the signal v contains a prediction of the output y, T units into the future. The new error term e' becomes,

$$e' = e - (\underline{\hat{P}}(s) - \hat{P}(s))u \tag{6.7}$$

where

$$\underline{\hat{P}}(s) = \frac{1}{\tau s + 1} \tag{6.8}$$

where the hat on top of the $P(s)$ denotes the model of the plant and the underline denotes the plant without the embedded delay. Assuming the model is perfectly matched, $\hat{P}(s) = P(s)$, (6.7) becomes,

$$e' = r - \hat{P}(s)u \tag{6.9}$$

As a result, transcendental term does not appear in the closed loop transfer function,

$$G(s) = \frac{y(s)}{r(s)} = \frac{CP(s)}{1 + C\underline{\hat{P}}(s)} \tag{6.10}$$

In the real world, there will always be model mismatch and as a result the transfer function will change to:

$$G(s) = \frac{y(s)}{r(s)} = \frac{CP(s)}{1 + C\underline{\hat{P}}(s) - C\hat{P} + CP} \tag{6.11}$$

Thus, even in the face of model mismatches, it can be stated that the Smith predictors reduces the effects of time delays in the system. However, looking at the transfer function from the disturbance of the output,

$$G(s) = \frac{y(s)}{d(s)} = P(s) \left(1 - \frac{CP(s)}{1 + C\underline{\hat{P}}(s)} \right) \tag{6.12}$$

88

it is seen the poles of this transfer function are zeros of $(1 + C\hat{\underline{P}}(s))$ and the poles of $P(s)$. Therefore, the Smith predictor control is valid only for stable plants [24].

6.3.2 Observer-Based Design

Watanabe and Ito proposed an observer for a linear feedback control law of multivariable systems with multiple delays in controls and outputs [96]. This observer based design overcomes the fact that the Smith predictors ignore the initial state of both the process and the delay element. Therefore, if the initial condition is not zero, the actual delayed output of the process cannot be exactly predicted. They proposed an observer, which can estimate the linear function of the state of the system from controls and outputs is proposed in their paper. Tarn and Brady [97] also used the observer-based design for the control of a time-delayed teleoperation system.

6.3.3 Sliding-Mode and Impedance Control

Park and Cho [98] developed a sliding-mode controller for bilateral teleoperation with variable time delay. They used a sliding-mode controller for the slave side and an impedance controller for the master side. Since sliding mode controllers are robust to variations in model parameters, they proposed to use such a controller under the influence of variable time delays. This modification adjusted the nonlinear gain to compensate for the effects of the variable time delays. The block diagram of the proposed control system is presented in the Figure 6.3.

Park and Cho also proposed impedance controllers on both sides of the teleoperation system as for the master and the slave systems. The simple sketch of the impedance controller for teleoperation systems is presented in the Figure 6.4.

6.4. Wave Variable Technique

The common shortcoming of the force-feedback teleoperation is the instability that the system undergoes when it experiences time delays in communications between the master and the slave. The magnitude of this time delay could be in the order of seconds, minutes, hours or days depending on the nature of teleoperation. This problem has been studied by many researchers, but Anderson and Spong were perhaps the first to use the wave variable method to control bilateral controllers [20]. Also, Niemeyer and Slotine [21], and Munir and Book [23-25] have implemented this method to teleoperation systems. Current studies are on

89

the variable time-delayed teleoperation [23-28]. Although the wave variable technique guarantees stability for the constant time delayed teleoperation, the system experiences instability when the time delay varies.

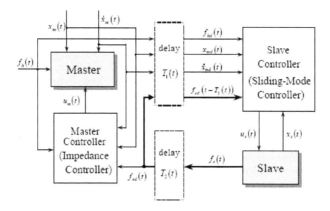

Figure 6.3. Sliding-mode / impedance controller block diagram
for bilateral teleoperation systems [98]

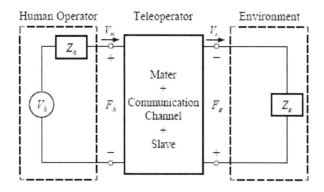

Figure 6.4. Basic sketch of impedance controller (Z being the impedance term)
for bilateral teleoperation systems [98]

The block diagram in Figure 6.5 below presents the wave variable technique in terms of the scattering transformation – a mapping between the velocity and force signals, and the wave variables [22].

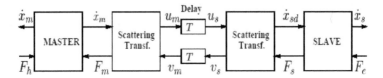

Figure 6.5. Scattering transformation for teleoperation with constant time delay [22]

This transformation using the notation in [23] is described as follows:

$$u_s = \frac{1}{\sqrt{2b}}(b\dot{x}_{sd} + F_s); \quad u_m = \frac{1}{\sqrt{2b}}(b\dot{x}_m + F_m)$$

$$\text{(6.13)}$$

$$v_s = \frac{1}{\sqrt{2b}}(b\dot{x}_{sd} - F_s); \quad v_m = \frac{1}{\sqrt{2b}}(b\dot{x}_m - F_m)$$

where $\dot{x}_m$ and $\dot{x}_s$ are the respective velocities of the master and the slave,

F_h is the torque applied by the operator,

F_e is the torque applied externally on the remote system,

F_m is the force reflected back to the master from the slave robot,

F_s is the force information sent from the slave to master,

$\dot{x}_{sd}$ is the velocity derived from the scattering transformation at the slave side,

u and v define the wave variables,

b is the wave impedance.

The power, P_{in}, entering a system can be defined as the scalar product between the input vector x and the output vector y. Such a system is defined to be passive if and only if the following holds:

$$\int_0^t P_{in}(t)d\tau = \int_0^t x^T y d\tau \geq E_{store}(t) - E_{store}(0)$$

$$\text{(6.14)}$$

where $E(t)$ is the energy stored at time t and $E(0)$ is the initially stored energy. The power into the communication block at any time is given by

$$P_{in}(t) = \dot{x}_{md}(t)F_m(t) - \dot{x}_{sd}(t)F_s(t) \tag{6.15}$$

In the case of the constant communications delay, the time delay T is constant,

$$u_s(t) = u_m(t-T); \quad v_m(t) = v_s(t-T) \tag{6.16}$$

Substituting these equations into (6.15), and assuming that the initial energy is zero, the total energy E stored in communications during the signal transmission between master and slave is found as

$$E = \int_0^t P_{in}(\tau)d\tau = \int_0^t (\dot{x}_{md}(\tau)F_m(\tau) - \dot{x}_{sd}(\tau)F_s(\tau))d\tau$$

$$= \frac{1}{2}\int_0^t (u_m^T(\tau)u_m(\tau) - v_m^T(\tau)v_m(\tau) + v_s^T(\tau)v_s(\tau) - u_s^T(\tau)u_s(\tau))d\tau \tag{6.17}$$

$$= \frac{1}{2}\int_{t-T}^t (u_m^T(\tau)u_m(\tau) + v_s^T(\tau)v_s(\tau))d\tau \geq 0$$

Therefore, the system is passive independent of the magnitude of the delay T. In other words, the time delay does not produce energy if the wave variable technique is used. Therefore, it guarantees stability for the constant time-delayed teleoperation.

For multi-DOF teleoperation systems, the inputs and outputs from the master and the slave are in vector form:

$$\underline{\dot{x}}_{sd} = \begin{bmatrix} \dot{x}_{sd} \\ \dot{y}_{sd} \\ \dot{z}_{sd} \end{bmatrix}; \quad \underline{\dot{x}}_m = \begin{bmatrix} \dot{x}_m \\ \dot{y}_m \\ \dot{z}_m \end{bmatrix}; \quad \underline{F}_s = \begin{bmatrix} F_s^x \\ F_s^y \\ F_s^z \end{bmatrix}; \quad \underline{F}_m = \begin{bmatrix} F_m^x \\ F_m^y \\ F_m^z \end{bmatrix} \tag{6.18}$$

These inputs and outputs from the master and the slave subsystems are transformed to wave variables using the B matrix for the multi-DOF case. The wave impedance matrix, B, is selected to be uncoupled as shown below:

92

$$B = \begin{bmatrix} b_x & 0 & 0 \\ 0 & b_y & 0 \\ 0 & 0 & b_z \end{bmatrix} \tag{6.19}$$

Munir and Book [23] wrote the wave transformation relation of equations in (6.13) in matrix notation to generalize it to multi-DOF systems as follows:

$$\underline{u}_s = A_w \dot{\underline{x}}_{sd} + B_w \underline{F}_s$$

$$\underline{u}_m = A_w \dot{\underline{x}}_m + B_w \underline{F}_m$$

$$\tag{6.20}$$

$$\underline{v}_s = C_w \dot{\underline{x}}_{sd} - D_w \underline{F}_s$$

$$\underline{v}_m = C_w \dot{\underline{x}}_m - D_w \underline{F}_m$$

where A_w, B_w, C_w, D_w, $B \in R^{nxn}$ (are nxn matrices);

$\underline{u}_s$, $\underline{u}_m$, $\underline{v}_s$, $\underline{v}_m$, $\dot{\underline{x}}_{sd}$, $\dot{\underline{x}}_m$, $\underline{F}_s$, $\underline{F}_m \in R^n$ (are nx1 vectors).

A_w, B_w, C_w and D_w are the scaling matrices and n is the degree of freedom of the teleoperation system. In the equations given earlier, $n=3$. Scaling matrices are determined using the impedance matrix (B), as follows:

$$A_w = \frac{\sqrt{2B}}{2}, \quad B_w = \frac{\sqrt{2B}}{2} (B^{-1}) \tag{6.21}$$

where usually C_w is selected to be the same as A_w, and D_w is selected to be the same as B_w [23]. Alise et al. have studied the formation of these matrices for multiple-degree-of-freedom teleoperation systems [99].

The customary formation of the wave variable technique is illustrated in Figure 6.5. The controller on the slave side is usually a velocity controller. A representation of the basic velocity controller block is shown in Figure 6.6. The velocity error is calculated in Cartesian space and then translated into joint velocity errors using the inverse of the Jacobian matrix, J. A general type of proportional-integral control is applied to calculate the driving torque for each joint. The N in the block diagram of Figure 6.6 represents the feedforward torque input to counteract the centrifugal, Coriolis and gravitational forces. Calculation of the errors in Cartesian space enables the usage of unlike masters and slaves.

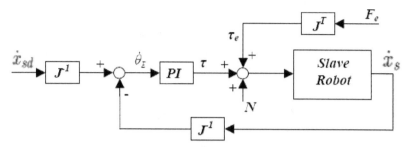

Figure 6.6. Customary velocity control of the slave manipulator

The system dynamics is written for teleoperation with the customary wave variable control as:

$$M_m \ddot{x}_m + B_m \dot{x}_m = F_h - F_m$$

$$M_s \ddot{x}_s + B_{s1} \dot{x}_s = F_s - F_e$$

(6.22)

where M_m and M_s are the respective inertias of the master and the slave, and B_m and B_{s1} are the master and slave damping respectively. The slave force can be formulated as;

$$F_s(t) = K_I \int_0^t \left(J^{-1} \dot{x}_{sd} - J^{-1} \dot{x}_s \right) ds + K_v \left(J^{-1} \dot{x}_{sd} - J^{-1} \dot{x}_s \right)$$

(6.23)

where K_I and K_v are the integral and proportional gain respectively.

This controller produces acceptable system response when the wave variable technique is active but the communication is never lost between the master and the slave. When the communication is lost for limited periods, an offset between the master and slave position tracking occurs.

94

6.4.1 Position Drift Compensator for Wave Variable Technique

A feedforward position demand is proposed to modify the wave variable technique to compensate for the offsets mentioned above. This demand is sent from the master system directly to the slave without integrating it in the scattering transform. This modification does not include a force feedforward component as in [100] because no drifts have been observed between the slave and the master force information in experimental studies [101]. The block diagram of the proposed algorithm is given in Figure 6.7.

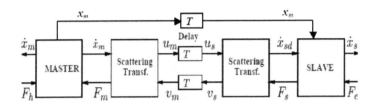

Figure 6.7. Offset compensation for the wave variable technique

The slave controller block diagram is also modified to comply with the new setting of the wave variable technique. As observed in Figure 6.8, the position error is calculated in the joint space. The motion demand from the master received in Cartesian space is transformed into the joint space by using the inverse of the Jacobian, J, and the inverse kinematics, IK. Later, the demand in joint space is compared to the joint sensor readings to form joint motion errors. This type of controller is of course feasible for those manipulators for which the inverse kinematics solutions are easy to obtain. Fortunately, almost all of the industrial manipulators are of this kind [102].

After the modification to the wave variable technique, the system dynamics is written as

$$M_s \ddot{x}_s + B_{s1} \dot{x}_s = F_s - F_e ; \quad M_m \ddot{x}_m + B_m \dot{x}_m = F_h - F_m \tag{6.24}$$

where

$$F_s = K_d \left(J^{-1} \dot{x}_{sd} - J^{-1} \dot{x}_s \right) + K_p \left(IK(x_m(t-T)) - IK(x_s) \right) \tag{6.25}$$

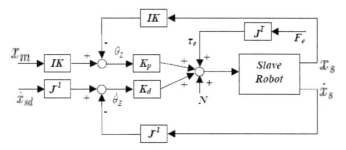

Figure 6.8. Modified controller for the slave system for the wave variable technique with position feedforward component

The control gains used in customary wave variable slave controller may be applied to the modified version. Therefore, the modified controller's gain magnitudes of K_d and K_p may be selected as equal to the magnitude of K_v and K_l. If the controller is considered as a velocity controller, this controller can be named as a PI controller. If the controller is considered as a position controller, the controller can be called a PD controller. In both cases, the architecture of the controller stays the same. Therefore, same control parameters can be used as described above.

6.4.2 Wave Variable Technique with Adaptive Gain

The block diagram in Figure 6.9 shows the modification described in [26] for the wave variable method for the variable time-delayed teleoperation.

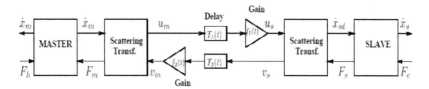

Figure 6.9. Wave variable technique with adaptive gain component for teleoperation systems experiencing variable time delays [26]

Time varying delay modifies the transmission equations to:

$$u_s(t) = u_m(t - T_1(t))$$

$$(6.26)$$

$$v_m(t) = v_s(t - T_2(t))$$

where, $T_1(t)$ is the variable time delay in the path from the master to the slave and $T_2(t)$ is the variable time delay in the path from the slave to the master. In [26], it is assumed that the frequency of change in time delays remain limited:

$$\frac{dT_i}{d\tau} < 1; \quad i = 1, 2 \tag{6.27}$$

Substituting the modified transmission equations to the equation for the total energy stored in the communication line (6.17)

$$
E = \frac{1}{2}\left[\int_{-T_1(t)} u_m^T(\tau)u_m(\tau)d\tau + \int_{-T_2(t)} v_s^T(\tau)v_s(\tau)d\tau \right.
$$

$$
- \int_0^{-T_1(t)} \frac{T_1'(\sigma)}{1 - T_1'(\sigma)} u_m^T(\sigma)u_m(\sigma)d\sigma \tag{6.28}
$$

$$
\left. - \int_0^{-T_2(t)} \frac{T_2'(\sigma)}{1 - T_2'(\sigma)} v_s^T(\sigma)v_s(\sigma)d\sigma \right]
$$

where $\sigma = \tau - T_i(\tau) = g_i(\tau)$ and $T_i'(\sigma) = \dfrac{dT_i}{d\tau}\bigg|_{\tau = g^{-1}(\sigma)}$.

The last two terms in (6.28) show that passivity cannot be guaranteed for variable time-delayed teleoperation. In the modified wave variable method, shown in Figure 6.6, a time varying gain f_i is inserted after the time varying delay block. Therefore, the new transmission equation becomes:

$$u_s(t) = f_1(t)u_m(t - T_1(t))$$

$$\tag{6.29}$$

$$v_m(t) = f_2(t)v_s(t - T_2(t))$$

The total energy stored can be re-written using the new transmission equations as

$$E = \frac{1}{2}\left[\int_{-T_1(t)} u_m^T(\tau)u_m(\tau)d\tau + \int_{-T_2(t)} v_s^T(\tau)v_s(\tau)d\tau \right.$$

$$- \int_0^{-T_1(t)} \frac{1-T_1'(\sigma)-f_1^2}{1-T_1'(\sigma)}u_m^T(\sigma)u_m(\sigma)d\sigma \qquad (6.30)$$

$$\left. - \int_0^{-T_2(t)} \frac{1-T_2'(\sigma)-f_2^2}{1-T_2'(\sigma)}v_s^T(\sigma)v_s(\sigma)d\sigma \right]$$

If f_i is selected that $f_i^2 = 1-T_i'$ for the total energy stored equation, the last two terms of the equation are eliminated and it can be said that the system is passive. In fact, the variable time-delayed system is considered to be passive if f_i satisfies the following condition [26]:

$$f_i^2 \leq 1 - \frac{dT_i}{dt}; \; i = 1, 2 \qquad (6.31)$$

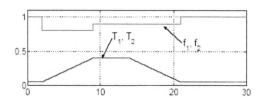

Figure 6.10. Change of the variable gain due to the change in time delay [26]

The change in the variable gain due to the variable time delay is computed using the relation described in Figure 6.10 [26]. This variable gain is called adaptive since it adapts itself with respect to the change in time delays.

6.4.3 Stability Analysis for Wave Variable Technique

Passivity is proven in (6.17) for customary wave variable technique and the communication line is also shown to be passive by selecting adaptive gains as shown in (6.31). The passivity guarantees stability for time-delayed teleoperation systems. Stability analysis can also be extended to check for Lyapunov stability for teleoperation systems using certain assumptions. Slave force F_s in (6.23) is rewritten for general teleoperation case as;

$$F_s(t) = K_s \int_0^t (\dot{x}_{sd} - \dot{x}_s) ds + B_{s2}(\dot{x}_{sd} - \dot{x}_s) \qquad (6.32)$$

where the control gains $K_s = K_I = K_p$ and $B_{s2} = K_v = K_d$. The assumptions [26, 103] used in the stability analysis is listed as: (1) Teleoperator and the environment are modeled as passive systems, (2) Teleoperator and the environmental forces are bounded by known functions of the master and the slave velocities respectively, (3) All signals belong to $\mathcal{L}_{2e}$, the extended $\mathcal{L}_2$ space, and (4) Initial velocities of the master and the slave must be equal to zero. Positive definite Lyapunov function for the system is selected as:

$$V = \frac{1}{2}\left\{M_m \dot{x}_m^2 + M_s \dot{x}_s^2 + K_s(x_{sd} - x_s)^2\right\} + \int_0^t (F_e \dot{x}_s - F_h \dot{x}_m) d\tau + \int_0^t (F_m \dot{x}_m - F_s \dot{x}_{sd}) d\tau \quad (6.33)$$

In the assumptions, it is stated that the human teleoperators and the slave environment are passive. As a result of this:

$$\int_0^t F_e \dot{x}_s d\tau \geq 0$$

$$-\int_0^t F_h \dot{x}_m d\tau \geq 0 \qquad (6.34)$$

The wave variable technique is shown to ensure passivity for teleoperation systems with time delays in (6.17) and (6.31). Therefore,

$$\int_0^t (F_m \dot{x}_m - F_s \dot{x}_{sd}) d\tau = \frac{1}{2}\int_{-T}^t (u_m^2 + v_s^2) d\tau \geq 0 \qquad (6.35)$$

Observing that the first term of the Lyapunov function and the terms presented in (6.34) and (6.35) are always positive, it can be said that the Lyapunov function is positive-definite. The derivative of this function is given as:

$$
\begin{aligned}
\dot{V} &= M_m \dot{x}_m \ddot{x}_m + M_s \dot{x}_s \ddot{x}_s + K\left(x_{sd} - x_s\right)\left(\dot{x}_m - \dot{x}_s\right) \\
&\quad + F_m \dot{x}_m - F_s \dot{x}_{sd} + F_e \dot{x}_s - F_h \dot{x}_m \\
&= \dot{x}_m\left(-B_m \dot{x}_m + F_h - F_m\right) + \dot{x}_s\left(-B_{s1}\dot{x}_s + F_s - F_e\right) \\
&\quad + K_s\left(x_{sd} - x_s\right)\left(\dot{x}_{sd} - \dot{x}_s\right) + \left(F_m - F_h\right)\dot{x}_m - F_s \dot{x}_{sd} + F_e \dot{x}_s \\
&= -B_m \dot{x}_m^2 - B_{s1}\dot{x}_s^2 + \left(\dot{x}_{sd} - \Delta v\right)F_s + \left(F_s - B_{s2}\Delta v\right)\Delta v - F_s \dot{x}_{sd} \\
&= -B_m \dot{x}_m^2 - B_{s1}\dot{x}_s^2 - B_{s2}\Delta v^2 \le 0
\end{aligned}
\tag{6.36}
$$

In (6.36), $\Delta v = \dot{x}_{sd} - \dot{x}_s$. The system can be said stable in the sense of Lyapunov as the derivative of the Lyapunov function is negative-semidefinite.

6.5. Position/Force Controllers in Teleoperation

Although fault tolerance secures the telemanipulation to continue for failures in motors, sensors, links and processors, it cannot yet secure the system when the system experiences communication losses. Communications line is often a source of failure in teleoperation systems. Communication may be lost completely, or the speed or quality of communication may fluctuate as observed in internet- or space-based communications. Heartbeat signals to detect the loss of communication is used in BOA II: Asbestos Pipe-Insulation Removal Robot System [104] and may be used in teleoperation systems to detect the loss of communication. The critical issue is to secure the system after the communication loss is detected. If such a failure occurs, the slave (remote robot or vehicle) is required to remain stable and not to damage itself or the environment by exerting excessive and uncontrolled forces.

In this study, position/force control algorithms are proposed as a parallel control algorithm that is to be activated at a time of communication loss. The reason of using these algorithms is that not only the velocity or the position of the slave side is monitored but also the force it is applying is controlled. While the communication is active, the human operator at the slave side accomplishes the force monitoring by the help of the force feedback signals. As the communication line or in other words the control over the force on the slave is lost, an automatic force control is required. This is where the position/force controller utilized. Researchers have been using position/force controllers in robotics application where the robot interacts with the environment. Some of the widely employed position/force control algorithms are stiffness, impedance, admittance, hybrid position/force, and hybrid impedance controllers [105]. Among these, hybrid position/force controller and the admittance controller is examined in this study.

6.5.1. Admittance Controller

Admittance control tracks not only the position trajectory but also the force trajectory. A pure position controller works on the principle of rejecting disturbance forces while following a reference motion. Instead of rejecting it, admittance control using a force compensator complies with the environmental interaction and reacts to contact forces by modifying the reference motion trajectory [106]. The mechanical admittance is defined by the equation below.

$$\dot{X}(t) = AF(t) \tag{6.37}$$

This equation is written in the s domain as

$$X(s) = K(s)F(s) \tag{6.38}$$

where

$$K(s) = \frac{1}{s}A \tag{6.39}$$

In (6.37), (6.38) and (6.39), and in Figure 6.11, X and $\dot{X}$ are the position and velocity vectors of the end-effector, and A is the admittance matrix. Figure 6.11 shows the schematic representation of a customary admittance control scheme.

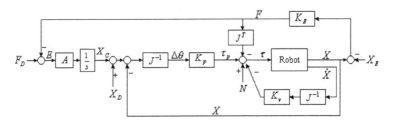

Figure 6.11. Customary admittance control [106]

The admittance matrix A relates the force error vector E ($E = F_D - F$) to the required modification in the end-effector velocity vector. This leads to the following additive modification on the reference trajectory:

101

$$X_c = \int A(F_D - F)dt \qquad (6.40)$$

Usually the admittance term, A, is not selected as a constant. It involves a variable matrix such as:

$$A(s) = k_d s^2 + k_p s + k_i \qquad (6.41)$$

which then results in the following PID force compensator when the (6.39) is applied:

$$K(s) = \frac{1}{s} \cdot A(s) = k_d s + k_p + \frac{k_i}{s} \qquad (6.42)$$

6.5.2. Hybrid Position/Force Controller

Hybrid position/force controllers combine position and force information into a single control signal to move the end-effector in nondeterministic environments [107]. Separate controllers process the position and force information independently so that the controller designer can take advantage of well-known control techniques for each of them. The outcomes of these controllers are then combined only at the final stage when both have been converted to joint torques. Figure 6.12 shows the application of the hybrid position/force control scheme as a block diagram.

In Figure 6.12, $S = diag(s_j)$ $(j = 1...n)$ is called the compliance selection matrix, where n represents the degrees of freedom. The matrix S determines the subspaces in which force or position are to be controlled, and s_j is selected as either 1 or 0. When $s_j = 0$, force control must be used in the jth direction of the Cartesian space; otherwise, position control must be used in that direction. Depending on the required task, the S matrix can be constant, or it can change in time according to the varying gradient of the task surface and the path followed on it.

For each task configuration, a generalized surface can be defined with position constraints along the normals to this surface and force constraints along the tangents, which means that the end-effector can not move along the normals (into the surface), and can not cause reaction forces to arise along the tangents of the surface. These two types of constraints partition the freedom directions of possible end-effector motions into two orthogonal sets

102

along which either position or force control must be used. Utilizing this partitioning, the S matrix is formed appropriately in accordance with the task requirements. In this control scheme, the command torque is calculated as:

$$\tau = \tau_p + \tau_f \qquad (6.43)$$

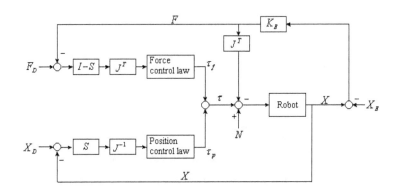

Figure 6.12. Customary hybrid position/force control [107]

τ_p and τ_f are the command torques created by position and force subspaces, respectively. In this way, position control and force control are decoupled. The control laws for each can be designed independently, so that the different control performance requirements for the desired position and force trajectory trackings are simultaneously realized. In general, it so happens that PD action is satisfactory for position control, and PI action is satisfactory for force control [108]. This is because for the position control, a faster response is more desirable, and for the force control a smaller error is more preferable.

6.5.3 Modification to Position/Force Controllers

Admittance and hybrid position/force control formulations use the assumption that the error between the position demand and the actual position in Cartesian space is small. Therefore, it can be transformed into the joint space using the approximation in (6.44).

$$(\theta_{ref} - \theta) \approx J^{-1}(X_{ref} - X) \qquad (6.44)$$

103

This assumption does not hold if there is a big enough time delay or communication loss during telemanipulation and the robot loses the track of its Cartesian coordinates. With the first command received the error range becomes unacceptable for this assumption.

Dede and Ozgoren [109] introduced a modification to this algorithm that provides a solution that does not use the assumption mentioned above. The modification is on exact calculation of the error in joint space. Therefore, both the position demand and the actual position measured in Cartesian space are transformed to the joint space using inverse kinematics (*IK*) as shown in (6.45). Usually the actual positions of the joints are received from the joint sensors in joint space. Then, the reference trajectory and the actual position can be compared in joint space without any assumptions. This solution is valid for the manipulators that have inverse kinematics solutions. In fact, almost all of the industrial manipulators are of this kind [102, 110].

$$(\theta_{ref} - \theta) = IK(X_{ref}) - IK(X) \qquad\qquad (6.45)$$

6.5.3.1 Modified Admittance Controller

The block diagram of the modified admittance controller is presented in Figure 6.13. As it can be observed, the modification is introduced in the inner position control loop where the error is calculated.

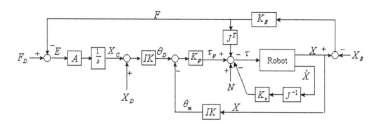

Figure 6.13. Modified admittance control [108]

Position feedback of the end-effector is changed to joint position feedback by inverse kinematics "*IK*" in the modified scheme. The inverse kinematics solutions can be achieved easily by using the methodology introduced in [102]. Besides, in a real time application, position feedback is received directly from the joint transducers. Therefore, it is sufficient to employ inverse kinematics only for the reference position $X_{ref} = X_C + X_D$ defined in the Cartesian space.

6.5.3.2 Modified Hybrid Position/Force Controller

Hybrid control scheme is also modified to make the necessary comparisons in the joint space and not in the Cartesian space as shown in Figure 6.14.

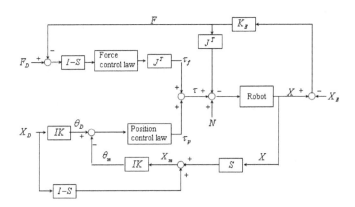

Figure 6.14. Modified hybrid position/force control [109]

This modified version of the hybrid control does not use the selection matrix, S, after the comparison is made in the Cartesian space for the position control subspace. The selection matrix is used to take the measured positions along the directions to be position controlled as they are and modify the measured position along the direction to be force controlled to the desired position along that direction. This makes the position error along the direction to be force controlled equal to zero, which means that the controller working in the position control subspace will not try to monitor the position along the force control direction.

After these modifications on the measured position, X, the modified measured position, X_m, is transformed to the joint space to calculate the modified measured joint angles, θ_m, using the inverse kinematics equations. Desired position vector is also transformed to the joint space using the inverse kinematics equations. As a result of this, the comparison is made in the joint space and the outcome is fed into the position controller.

Although modified hybrid position/force controller also provides satisfactory behavior when large errors in Cartesian space are observed, the control algorithm requires dual controls. In order to approach the manipulator to the surface to make contact, the algorithm requires a pure position controller. As soon as the contact is formed, algorithm has to switch from pure position controller to hybrid position/force controller. In a teleoperation system that experiences time delays, this type of architecture would cause possible instabilities and chattering. Therefore, in this study the tests are carried out using the modified admittance controller, which is the most suitable controller for time-delayed teleoperation by its architecture.

CHAPTER VII

NUMERICAL SIMULATIONS

7.1. Introduction

Communications lines of teleoperation systems are not always dependable. The system can experience time delays or communication losses during the telemanipulation. As a result of this, teleoperation systems experience instabilities. The wave variable technique and its versions were proposed to resolve this problem in the previous chapter. In this chapter, these teleoperation controllers are examined in numerical simulations.

When communication losses are observed, the slave system or its environment can also be damaged. The reason for this is that the slave interacts with the environment as it carries out its mission. Most of the time, this interaction requires the application of forces to the environment. When the communication loss is experienced, the slave may exert excessive amounts of forces to the environment which causes damages in the environment and itself. Customary motion control schemes in general do not provide satisfactory response for these types of applications. Implementation of a force controller may provide a more satisfactory solution. Another solution is utilization of parallel position/force controllers. These controllers are explained in the previous chapter. In this chapter, a simulation study is also conducted for the performance of position/force controllers in teleoperation systems.

The next two sections describe the single- and multi-DOF teleoperation system models that are used in simulations. In these sections, the models of three subsystems, which are the master, slave and the communications line, are explained in detail. Matlab$^{©}$ Simulink blocks and VRRP method are used to build these models. The first numerical simulation tests are carried out for a single-DOF teleoperation. The wave variable technique is tested as the teleoperation controller in these tests. Effect of wave impedance "b" on the telemanipulation is also investigated using the results of this simulation study. The time delays that cause instabilities are kept constant in the single-DOF tests.

Teleoperation model is then extended to a multi-DOF system. Initially, constant time-delayed teleoperation is studied in the simulations of a multi-DOF system. The customary wave variable technique is used in the tests to stabilize the system. Later, the system is exposed to variable time delays in the communications line. The performance of the customary wave variable technique is compared to the version when adaptive gain component is included. Finally, two position/force controllers are discussed and the simulation results illustrating the performance of the admittance controller are presented as applied on a SCARA manipulator.

7.2. Single-Degree-of-Freedom Teleoperation System Model

The teleoperation system mainly has two subsystems other than the communications line: The master controller, which is, in this case, a one-degree-of-freedom (DOF) joystick, and the slave robot, which is modeled as a one-DOF slider. These two subsystems are modeled in Matlab© using the Simmechanics blocks of Simulink. The subsystems are shown in Figures 7.1 and 7.2.

The torque input applied by the operator on the joystick, denoted by "Joy_Out" in the block diagram (Figure 7.1), is fed to the joint actuator of the joystick with the force feedback information from the slave robot and the joystick spring dynamics output, "Torque of Spring." The "Spring&Damper" block is used to model a spring system to move the joystick to the null position when there is no other torque applied to it. It is composed of simple Simulink blocks that multiply the position and velocity feedback with certain gains to make the block act as a spring-damper system. Force feedback information from the slave is either sent while there is a time delay by "Slave_FF" or while there is no time delay by "Force_FB," which is switched by the "Time_Dly" switch input generated from the main window. The rest of the blocks of Figure 7.1 show the blocks from Simmechanics library of Simulink to model the kinematics and dynamics of the joystick. The Simmechanics blocks that are used to develop the master and the slave robot are defined in Table 3.1.

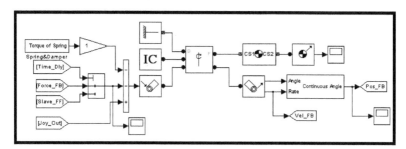

Figure 7.1. Master (joystick) subsystem window

Figure 7.2 shows the Simulink window of the modeled slave robot. The kinematics and dynamics of the robot is also modeled with the Simmechanics library of Simulink.

Different than the master, the slave has one prismatic joint, which enables it to work like a slider mechanism as a one degree of freedom system. The slave robot simply takes the velocity command from the master, "Slave_V_W," if there is a time delay or it is switched to take the velocity command from the master output directly, "Pos_FB," by the help of the "Time_Dly" switch and compares it with its velocity feedback "Slave_V" to feed the necessary information to the PD controller. Also, it sends the necessary output to create the force information in means of proximity to the modeled wall, by the position of itself "Slave_P."

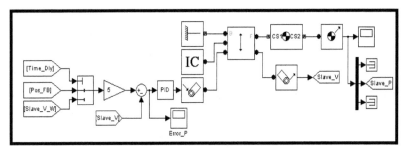

Figure 7.2. Slave sub-system window

Figure 7.3 shows the communication between the master and the slave. Force feedback information is created with the "FF Command" in Figure 7.3. "FF activation" block senses the contact of the slave robot with the environment and enables the force feedback information to be sent to the master by switching from zero input block to "FF Command" block. There are four other switching conditions to enable usage of the wave variable technique for the time delayed teleoperation. These switches are operated by the input "Wave_Vrb" generated from the main window. The rest of the blocks of the "Communication Line" block is to construct the wave variable method into the communication line between the master and the slave. The amount of time delay is set through the "Time Delay" blocks.

The main control window of teleoperation is shown in Figure 7.4. The subsystems are marked with "Master (Joystick)" and "Slave." The generation of time delay and the application of the wave variable technique to the communication line are in the "Communication Line" block of the main control window. The operator's interaction to apply torque to the joystick is accomplished through the joystick with the tag "Operator

Torque Input." The motion of the master robot (joystick) under the influence of the torque input from the operator and the force feedback provided from the slave robot is observed from the joystick with the tag "Actual Joystick Motion."

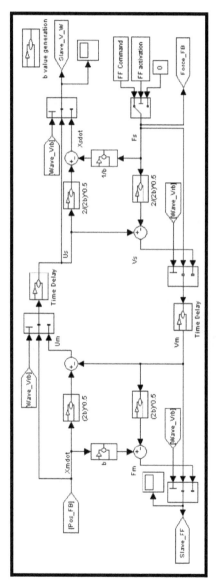

Figure 7.3. Communication line block window

The motion of the slave robot (slider) is observed from the slider on the main control window with the tag "Slider Motion." There are also two switches that appear on the main control window of teleoperation. The first one with the tag "Time Delay On/Off" is to enable the time delay on the communication line of the system. This switch generates an input, "Time_Dly," for the switching in the master and the slave robot. The second switch with the tag "Wave Variables On/Off" enables the application of wave variable technique to the system with constant time delays. This switch also generates an input, "Wave_Vrb," activating the wave variable technique in the "Communication Line" block.

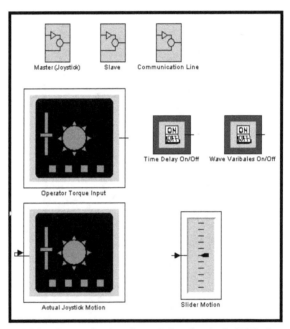

Figure 7.4. Main teleoperation interface window for single-DOF teleoperation

7.3. Multi-Degree-of-Freedom Teleoperation System Model

The multi-DOF teleoperation system has also two subsystems other than the communications line: The master controller, which is modeled as a three-DOF joystick, and the slave robot, which is a three-DOF Cartesian robot. The joystick is modeled as an uncoupled system in terms of its three DOF since this is the case for the actual gimbal-based joystick used in experiments detailed in the next chapter. The only difference between this

111

model and the actual joystick is that the joystick has two DOF instead of three. The two subsystems are modeled in Matlab[©] using the Simmechanics blocks of Simulink. Master and slave subsystems are shown in Figures 7.5 and 7.6.

Torque inputs applied by the operator on the joystick, denoted by "Joy_Out" in the block diagram (Figure 7.5), are fed into the joint actuators of the joystick with the force feedback information from the slave robot and the joystick spring dynamics output, "Torque of Spring." The "Spring&Damper" blocks are used to model a spring system to move the joystick to the null position when there is no other torque applied to it. It is composed of simple Simulink blocks that multiply the position and velocity feedback with certain gains to make the block act as a spring-damper system. Force feedback information from the slave is either sent (1) while there is a time delay by "Slave_FF" or (2) while there is no time delay by "Force_FB," which is activated by the "Time_Dly" switch input generated from the main window. The rest of the blocks of Figure 7.5 are the blocks from Simmechanics library of Simulink to model the kinematics and dynamics of the joystick.

Figure 7.6 shows the Simulink window of the modeled slave robot. The kinematics and dynamics of the robot is also modeled with the help of Simmechanics library. Different than the master, the slave has three prismatic joints, which enable it to work like a Cartesian robot with three-DOF. The slave robot receives the velocity command from the master through the "Slave_V_W" block if there is a time delay. If it is switched to receive the velocity command from the master output directly through the "Pos_FB" block with the help of "Time_Dly." Then, it compares the velocity input with velocity feedback of the slave through the "Slave_V" block to feed the necessary information to the PD controller. Also, it sends the necessary output to create the force information in relation to the proximity to the constraints, by comparing the position information in three-DOF, "Slave_P."

Figure 7.7 presents the communications protocol between the master and the slave. There are four switching conditions to enable usage of the wave variable technique for the time-delayed teleoperation. These switches are operated by the input "Wave_Vrb" generated from the main window. The rest of the blocks of the "Communication Line" block are to construct the wave variable method into the communication line between the master and the slave. The amount of time delay is set from the "Time Delay" blocks. Also in this block there is a switch between the constant time delays to the varying time delays which is operated by the input block "Varying Time Delay On/Off" in the main interface window in Figure 7.8. The switching from the wave variable technique to the modified method is accomplished in the "Vr/no Gain" blocks by enabling or disabling the adaptive gain, which is operated by the switch "Varying Gain on/Off" in the main interface window in Figure 7.8. All the calculations carried out in the communications line block are matrix based. The lines between the blocks carry information for three-DOF in vector format.

The main control window of teleoperation is shown in Figure 7.8. The subsystems are marked with "Master (Joystick)" and "Slave." The generation of time delay and the application of the wave variable technique to the communications line are in the "Communication Line" block of the main control window. The conversion of information to matrix and vector format is accomplished in a block named "Matrix conversions." The force information from the position of the slave is also created by this subsystem. The operator's interaction to apply torque to the joystick is incorporated through the joystick torque inputs "Joy_Out_1," "Joy_Out_2," and "Joy_Out_3" representing the y, x and z axes, respectively.

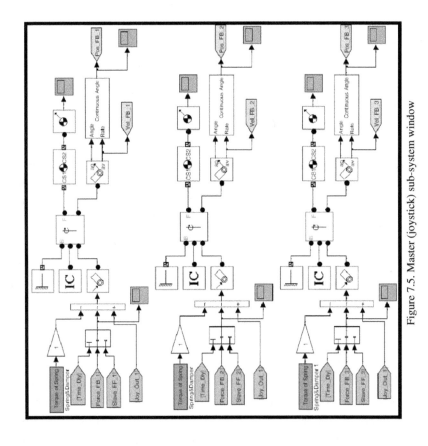

Figure 7.5. Master (joystick) sub-system window

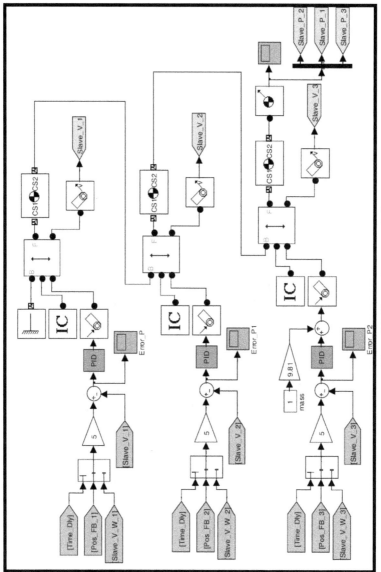

Figure 7.6. Slave sub-system window

114

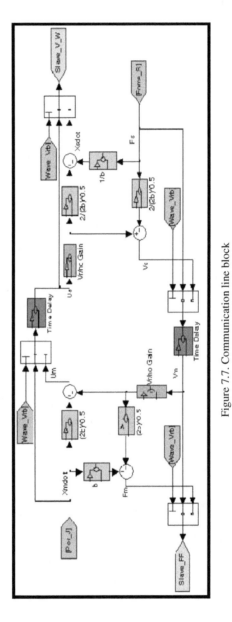

Figure 7.7. Communication line block

The motion of the slave robot (Cartesian robot) is observed from the scope in the "Slave" subsystem. There are also four switches that appear in the main control window of the teleoperation. The first one with a tag "Time Delay On/Off" is to enable the time delay on

115

the communications line of the system. This switch generates an input, "Time_Dly," to switch the time delay "on" or "off" in the master and the slave robot. The second switch with a tag "Wave Variables On/Off" enables application of the wave variable technique to the system with a constant time delay. This switch also generates an input called "Wave_Vrb" to accomplish the necessary switch in the "Communication Line" block. The third switch is to enable the varying time delay, which is denoted by "Varying Time Delay On/Off." This output from this switch changes the behavior of the delay from constant to varying. The fourth and the last one of the switches is the switch to enable the adaptive gain in the communication line to switch from the wave variable technique to the modified one. This switch is denoted by "Varying Gain On/Off" in Figure 7.8.

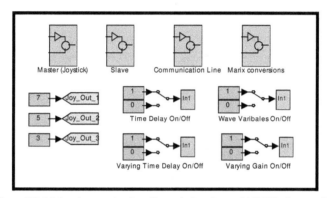

Figure 7.8. Main teleoperation interface window for multi-DOF teleoperation

7.4. Teleoperation Controller Simulation Results

Three types of simulation results are presented in this section. The first set of results is for the single degree-of-freedom constant time delayed teleoperation. In the second set, results for the multiple degree-of-freedom constant time delayed teleoperation are given. The last set of simulation results are for variable time-delayed teleoperation simulations.

7.4.1 Test Results for Single-DOF Teleoperation with Constant Time Delays

The first simulation carried out in this study models a communications line with no time delay. This simulation gives an idea of the ideal case where the two subsystems are coupled perfectly without any delays. The second set of simulations is carried out for time delays of 0.1, 0.2 and 0.5 second and in the absence of wave variable technique. This set of

116

simulations provides information on how time delays play a role in the stability of teleoperation. Then, the next set of simulations is carried out for time delays of 0.1, 0.2 and 0.5 second in the presence of wave variables to demonstrate stability in teleoperation. Finally, the simulation results for a 0.5 second delayed teleoperation having different wave impedance values are given. The results of these numerical simulations are also presented in [111, 112].

The task for each simulation is set as follows: The operator applies a steady torque to the master controller (joystick) to send a constant velocity command to the slave. The slave sliders proximity sensor is set to 50 inches. Therefore, as it reaches beyond the set value of 50 inches, the slave slider sends force information to the master with respect to the distance violated beyond the limit. During all this time, operator still exerts the constant torque to the joystick to make the slave slider move in the same direction. This type of operation is likely to cause an oscillatory motion about the constraint, which should be damped to a position just above the limiting value of 50 inches due to the steady operator torque input.

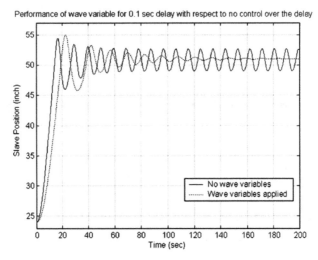

Figure 7.9. Effect of wave variable technique on a 0.1 second time-delayed teleoperation

Figures 7.9, 7.10, and 7.11 are presented to demonstrate the effect of wave variable technique on the stability of teleoperation. The solid lines on the plots represent the slave motion in the absence of wave variable technique for the communication between the master and the slave. The dashed line shows the slave response in the presence of wave variables. It

117

can be observed from the figures that when the wave variable technique is not activated, the slave motion oscillates without any damping to converge the motion to a steady state. As the wave variable technique is activated, the motion of the slave is dampened and therefore it converges to a point just above the limiting value of 50 inches.

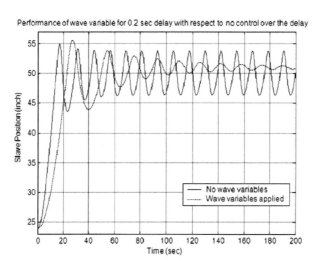

Figure 7.10. Effect of wave variable technique on a 0.2 second time-delayed teleoperation

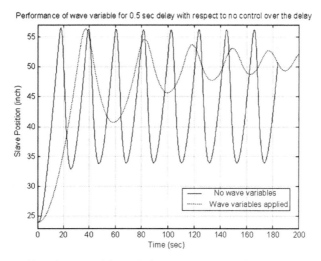

Figure 7.11. Effect of wave variable technique on a 0.5 second time-delayed teleoperation

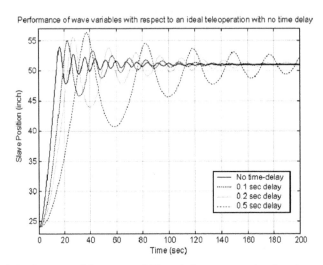

Figure 7.12. Performance of the wave variable technique compared to the teleoperation with no time delay

While guaranteeing the stability of the teleoperation with a constant time delay, the decrease in the manipulation speed caused by the application of wave variable technique can be observed from the Figures 7.9-11. Figure 7.12 illustrates a different collection of system responses ranging from no time delay to 0.1, 0.2 and 0.5 second of time delays.

It is observed from Figure 7.12 that as the time delay increases, the magnitude of overshoot increases but the manipulation speed decreases. Also an increase of time delay results in an increase in the time required to settle the slave slider.

Selecting the wave impedance for the wave variables controller one should be very careful not to make the teleoperation system unstable. Tuning of the wave impedance term, b, enables to change the characteristic of the teleoperation system as it can be observed from Figure 7.13. As the wave impedance term increased, it can be observed that the oscillation magnitude increases while the oscillation frequency decreases. This means that increase in the impedance term causes the manipulator to settle in steady state in longer time with bigger overshoots in transition state. Also, as another outcome of this simulation the wave impedance term should be selected in a range that will not cause instability. The range turned out to be between 400 and 800 for the system simulated.

119

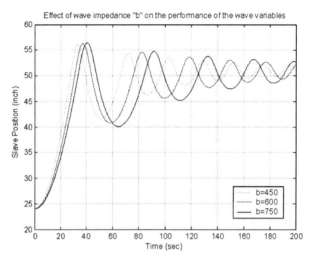

Figure 7.13. Performance variation of the wave variable technique with respect to the changes in the wave impedance

7.4.2 Test Results for Multi-DOF Teleoperation with Constant Time Delays

The first set of simulations is carried out on the multi-DOF teleoperation system for time delays of 0.1, 0.2 and 0.5 second and in the absence of the wave variable technique. This set of simulations will provide information on how the time delay plays a role in the stability of teleoperation and how oscillary motion increases with increasing time delays. The second set of simulations is carried out for time delays of 0.1, 0.2 and 0.5 second in the presence of wave variables to improve stability in teleoperation. The results of these numerical simulations are also presented in [113].

The scenario for these simulations assumes that the operator applies constant but differing amounts of torques to each of the three degrees of freedom of the joystick to send a constant velocity command vector to the slave (remote Cartesian robot). The slave robot's proximity sensors are set to 50 in for the x-axis, 30 in for the y-axis and 70 in for the z-axis. Therefore, as it goes beyond the limits, the slave robot sends force information in vector format to the master with magnitude proportional to the distance violated beyond the limits. During all this time, the operator still exerts constant amounts of torque to the joystick to make the slave robot move in the same direction. This type of operation is likely to cause an oscillatory motion about the constraint, which should be damped to a position just above the 50-meter limit due to the steady input torque provided by the operator.

Figures 7.14, 7.15, and 7.16 are presented to illustrate the effect of increasing time delays on the stability and oscillations of the manipulation. Slave motion oscillates about different limits for each degree of freedom, as it was set. It is noted that the magnitude of oscillations increases, as does the time delay.

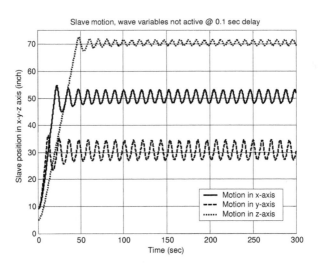

Figure 7.14. Effect of 0.1 second time delay on three-DOF teleoperation

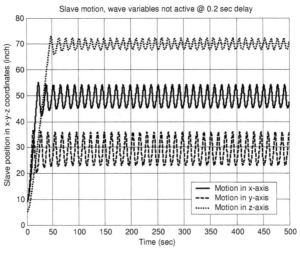

Figure 7.15. Effect of 0.2 second time delay on three-DOF teleoperation

121

It is observed from the Figures 7.14-16 that when the wave variable technique is not activated the slave motion oscillates without any damping. The Figures 7.17, 7.18, and 7.19 are presented to highlight the effect of the wave variables on the same system that was unstable under time delays of 0.1, 0.2, and 0.5 second, respectively.

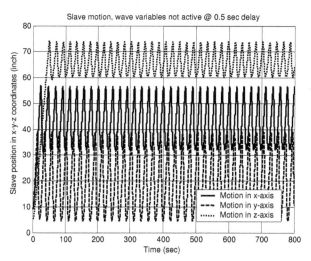

Figure 7.16. Effect of 0.5 second time delay on three-DOF teleoperation

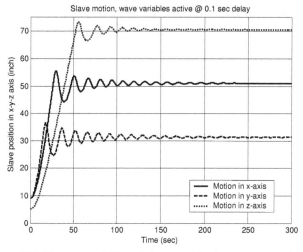

Figure 7.17. The wave variable technique used for 0.1 second time-delayed
three-DOF teleoperation

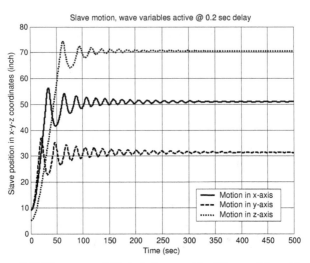

Figure 7.18. The wave variable technique used for 0.2 second time-delayed three-DOF teleoperation

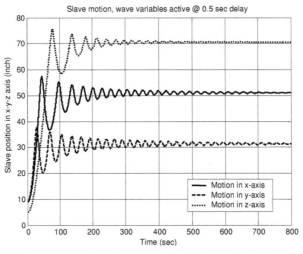

Figure 7.19. The wave variable technique used for 0.5 second time-delayed three-DOF teleoperation

As the wave variable technique is activated, the motion of the slave is dampened, and converged to a point just above the limiting value of 50, 30, 70 inches for x, y and z axes, respectively. In general, larger settling times are observed when higher time delays are modeled in communication lines. For instance, the 0.1 second time-delayed teleoperation settles in about 300 seconds whereas it is about 500 seconds for 0.2 second time delay, and 800 seconds for 0.5 second time delay; all corresponding to the same wave impedance.

7.4.3 Test Results for Multi-DOF Teleoperation with Variable Time Delays

The task defined for the constant time-delayed teleoperation is also used for the variable time-delayed teleoperation simulations. Figure 7.20 shows the case when there is variable time delay in the system and no wave variables control over it. Time delays in the communications line of the system range from 0.5 second delay to 0.1 second delay. The results of these numerical simulations are also presented in [114].

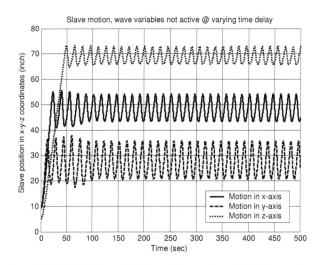

Figure 7.20. Effect of variable time delay on three-DOF teleoperation

Figure 7.21 shows the case when there is variable time delay and the wave variable technique is applied. It is observed from the figure that the system is still unstable under the influence of the wave variables for the variable time delay but the oscillations are damped with respect to the case where the wave variable technique is not applied.

Figure 7.22 shows the case when the modified wave variable technique is applied to the system with variable time delay. As shown in the figure, the adaptive gain placed just after the time delay in the communication lines as explained in the previous chapter guaranteed the stability for the variable time delay case.

124

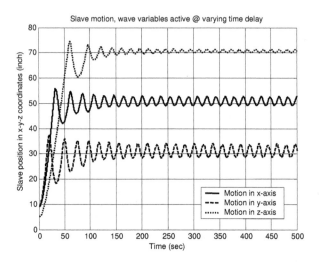

Figure 7.21. The wave variable technique used for the variable time-delayed three-DOF teleoperation

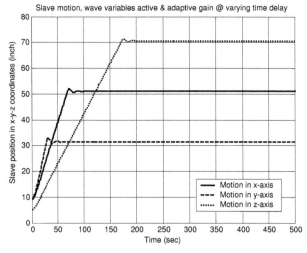

Figure 7.22. The wave variable technique with adaptive gain used for the variable time-delayed three-DOF teleoperation

7.5. Position/Force Controller Simulation Results

Position/force controllers are tested using the Epson SCARA manipulator model presented in Chapter V. The slave system is a four-DOF SCARA manipulator. Therefore, it

125

has limited workspace and requires mapping between the joystick motion and the slave motion. This mapping is accomplished between the Cartesian space positioning of both master and the slave [67]. The task is to trace surfaces by exerting controlled forces. The commands received in x- and y-axes are transmitted to the slave as Cartesian coordinate inputs. Then, the forces created during the telemanipulation as a result of interaction and surface friction are fed back to the master.

Interaction model between the end-effector and the surface is created using a planar contact. The material of the end-effector is selected to be lead with a modulus of elasticity of 36.5 GPa. The surface is assumed to be rigid.

The operator specifies the desired force to be applied to the environment through an input screen. This force profile can also be called the force trajectory. The interaction force between the surface and the end-effector as well as the surface friction force are created in the simulation environment. The friction forces in Cartesian space are then fed back into the servomotors of the joystick's respective axes. The friction forces are also sent back to the master as force feedback information.

In the simulations, independent joint control is used, which provides the flexibility of using different control algorithms for each joint. In the first set of simulations, Proportional-Derivative (PD) control was implemented for every joint. After tuning the PD parameters, the controller produced an acceptable error range when there was no contact with the surface. In order to control the force applied on the surface with this controller, the desired position trajectory was modified to penetrate into the surface so as to create the desired amount of contact force.

As expected, a pure position controller for the prismatic joint was not effective enough to follow the force trajectory while following the position trajectory. Another solution is to use dual controls. A position controller makes the end-effector approach the surface and create the contact. Then the control algorithm has to be switched to a pure force controller or to hybrid position/force controller to follow the force trajectory. This type of switching between the controls can cause instabilities and chattering in a teleoperation system where the communication can be delayed in an unacceptable amount.

One other possibility for a system that is required to follow both position and force trajectories is to use the modified admittance controller presented in the previous chapter. The following set of simulations is carried out by using an admittance controller for the prismatic joint and PD controller for the revolute joints. The PD control parameters were modified to

126

compensate for the disturbances created by the friction forces at revolute joints. The simulation results are also presented in [115].

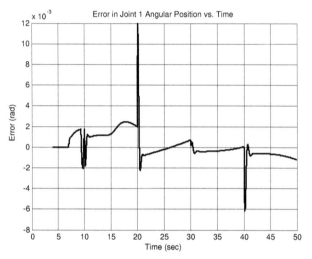

Figure 7.23. Angular position error in the first joint

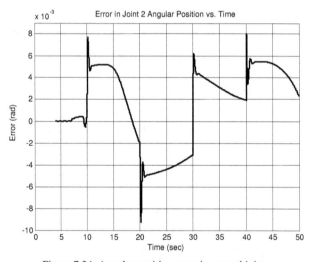

Figure 7.24. Angular position error in second joint

The task used in the simulations is to follow a square path inside the workspace of the manipulator by applying a constant force. The task was made more demanding by specifying the speed of the end-effector constant. It was expected to cause problems especially at the corners of the square where the end-effector is required to change its direction by 90°. The maximum amount of error is seen at the change of direction at 20, 30 and 40 seconds as illustrated in Figures 7.23 and 7.24.

The change of direction acts as a step input because of the design of the task. This can be observed clearly from the velocity response of the manipulator in Cartesian space in Figure 7.25.

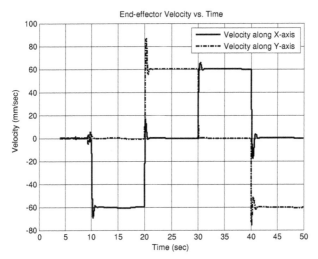

Figure 7.25 End-effector velocity in x- and y-axes

The square drawn by the end-effector also has the characteristics of transition states after each change of direction. The simulation creates an output for the lines drawn. The graph in Figure 7.26 is the output for the lines drawn on the surface.

The force applied to the surface is presented in Figure 7.27. It is observed that after an acceptable transition period, the contact force is kept stable at the designated amount without any overshoots. The transition state characteristics can be changed by modifying the admittance term of the controller. The position control law however is kept constant as a PD controller. It was not necessary to modify the position control parameters. The position was tracked in acceptable error range with the inner position loop of the admittance controller.

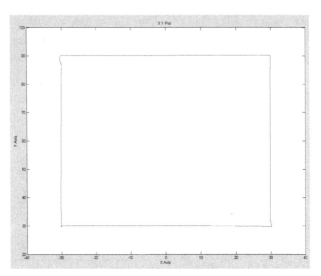

Figure 7.26. Lines drawn on the surface

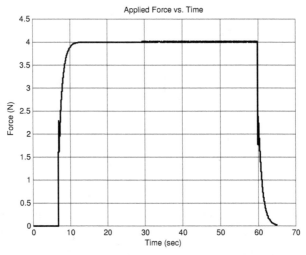

Figure 7.27. Force applied by the end-effector to the surface

The friction force created in the Cartesian space is presented in Figure 7.28. This information is to be transmitted to the servomotors of the joystick as torque inputs in each

129

axis. It is also observed from this figure that the friction force alters with the change of direction of motion.

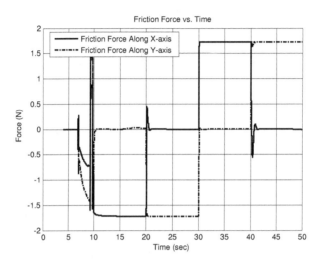

Figure 7.28. Friction force observed along each axis

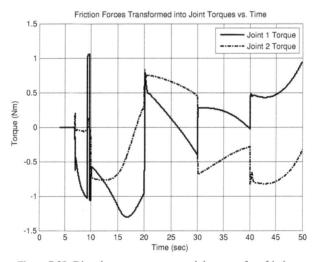

Figure 7.29. Disturbance torques created due to surface friction

Friction force information is then transformed into joint disturbance torques. It was necessary to have this transformation so that the friction forces created would affect the manipulator motion as expected. The disturbance torques are presented in Figure 7.29.

The tests show that the virtual slave manipulator follows the force trajectory within an acceptable error range and with acceptable transients while following the motion trajectory. It also creates the force feedback information for the operator to provide a feel of the remote environment.

7.6. Conclusions

In this chapter, modeling of single and multi degree-of-freedom teleoperation systems, and the simulation results of this teleoperation system for different tasks are presented. Although the main task remains the same for each simulation, changing the time delay and activating and deactivating the wave variables in simulations provided a better understanding of the necessity of the wave variable method in constant time delayed teleoperation.

As the wave variable technique enhances the stability of constant time delayed teleoperation, it is observed that the drawbacks of the wave variables are listed as an increase in the magnitude of the overshoot as the system response reaches the set point, and also the general decrease in manipulation speed. The decrease in manipulation speed and increase in the overshoot is observed more clearly as the time delay increases. Also, tuning of the wave impedance term for an optimum manipulation speed and overshoot is required.

The teleoperation system performance is then tested under variable time delays. The test results have shown that the customary wave variable technique is not sufficient to fully stabilize the system. A modified version of the wave variable technique with adaptive gains is tested for the same system. It is observed from the test results that this proposed new controller ensured stability for teleoperation systems with variable time delays.

A teleoperation system is required to remain stable while performing critical tasks even though it experiences communication or data losses. Wave variable technique cannot guarantee that the system is not harmed during these periods. Previously in this study, position/force controllers were proposed for use when the communication is lost. In that regard, it is observed that the selection of parallel position/force controllers for teleoperation applications becomes very crucial.

In this chapter, two position/force control algorithms used in many other robotics applications are examined for use in teleoperation. The main objective was to compare the control algorithms for stability in case of time delays or communication losses. The customary versions of the hybrid position/force and admittance controllers fail in teleoperation system applications. The reason for the failure is because of the assumption that the error between the demanded and the actual positions in Cartesian space is small for a finitely small amount of time. This assumption becomes invalid when the system experiences time delays or loss of communication.

This assumption is not necessary for the proposed, new versions of these controllers presented earlier when the comparison between the demanded and the actual signals is made in joint space. On the other hand, hybrid position/force control requires a pure position controller for switching from one to another while in transition from contact to no contact. Therefore, possible instabilities and chattering are foreseen in case of time delays or communication losses.

The proposed admittance controller does not require switching to a pure position controller regardless of the contact condition. The simulation results are also given to examine the performance of controllers in telemanipulation tasks. Finally, it is concluded that among the position/force controllers examined, admittance controller is seen as the best choice for teleoperation applications that may experience communications line problems and require interaction with the environment by applying forces.

CHAPTER VIII

LIMITED-WORKSPACE TELEOPERATION EXPERIMENTS

8.1. Introduction

Simulation studies presented in the previous chapter is extended to experimental studies with teleoperation test systems developed for this work. The master and slave systems that are used in these experiments were introduced in Chapter V. The teleoperation experimental setups are configured as combinations of these master and slave systems such that they form a limited-workspace teleoperation.

Teleoperation systems using serial or parallel slave manipulators with limited workspace are defined as limited-workspace teleoperation. Telemanipulation of an industrial robot arm is an example to this type of teleoperation. In this chapter, three types of limited-workspace teleoperation experiments are presented.

The first set of experiments is conducted for the identical master-slave teleoperation system. This setting does not require special mapping of the motion between the master and slave. The motion about each axis is fed through directly and independently. Two-DOF gimbal-based joystick and its virtual replica are used in these experiments.

Industrial arms are used for the slave systems in the second set of experiments. Master systems are the two-DOF gimbal-based master joystick and the Phantom Omni Device. The motion and force demands of this setup is calculated in Cartesian space and then translated into joint space. Another study is conducted for the commercial joystick and the Phantom Omni Device. In this study Phantom Omni Device is used as a slave industrial arm.

For both set of experiments, position tracking performance studies are conducted at constant time delays. Later, the systems are tested for variable time delays. The wave variable technique and its modifications proposed in Chapter VI are examined as a result of these tests for limited-workspace teleoperation systems.

The last set of experiments is conducted for the teleoperation system that is composed of the gimbal-based joystick and the Epson SCARA manipulator. The position/force control algorithms that are introduced in Chapter VI are examined in these experiments.

8.2 Identical Master-Slave Experiments

The two-DOF gimbal-based joystick is used as the master in these experiments. The joystick has uncoupled motions about the two axes due to its gimbal-based design. The slave

133

system is designed to be the replica of the master system. This indicates that the slave has the same workspace with the master and does not require mapping between the master and the slave motions. The master joint motion readings are directly sent as demands to the joint actuators of the slave. The slave is constructed as a virtual robot. Its virtual representation and system model is presented in Chapter V. The experimental setup used in identical master-slave teleoperation is shown in Figure 8.1.

The integration of the two subsystems of teleoperation is accomplished through the Matlab$^{©}$ Simulink environment. Galil motion control card (DMC 18x2) is used to drive the servomotor on the actual joystick and also to gather the encoder readings from each axis. A driver is developed as an interface in simulation environment to send and receive data with the control card. The simulation is then synchronized with the real-time clock to run the real-time tests. The sampling rate is selected to be 100 Hz. This is an acceptable rate for testing a system that can experience 0.1 to 0.5-second time delays.

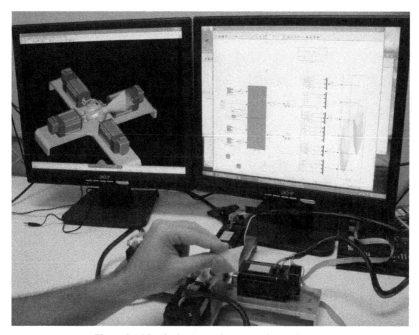

Figure 8.1 Identical master-slave experimental setup

8.2.1 Experimental Results with Constant Time Delays

In this experiment, the customary wave variable technique is utilized in the *x*-axis motion, and the wave variable technique with the position feedforward component for the *y*-axis motion. Since both axes are uncoupled, the motion along one axis does not affect the motion along the other axis. Hence, comparison of both controllers becomes possible in one experiment. The results of these experiments are also presented in [116, 117].

During the test, the scenario is set up such that the operator moves the joystick away from the home position and returns it to imitate a step input. The slave trying to follow these commands is steered towards an obstacle. As a result of this, force information is generated and sent back to the master. The first tested failure is the deactivation of the wave variable technique.

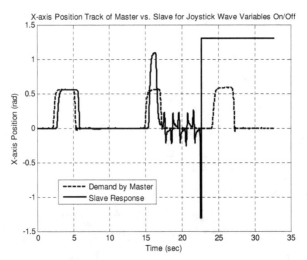

Figure 8.2. Position tracking performance of the identical master-slave teleoperation using the customary wave variable technique

The test started with the controller using the wave variable technique and experiencing 0.5 second time delay in the communication line. Until the 14th second, there were no explicit errors in position tracking of the slave. At the 14th second, wave variables are switched off and the data is transferred through the regular communication line. The system became unstable and the forces are observed at very high limits as seen in Figure 8.2. At the 23rd second, the wave variable technique is switched on and the system was stabilized

(Figure 8.3). Although the system was stabilized, *x*-axis position of the slave was observed to be at the mechanical stoppers of that axis. Therefore, the slave could not align its position trajectory with the master as observed in Figure 8.2.

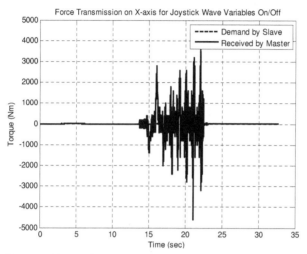

Figure 8.3. Force tracking performance of the identical master-slave e teleoperation using the customary wave variable technique

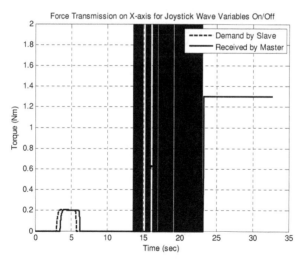

Figure 8.4. Force tracking performance of the identical master-slave teleoperation using the customary wave variable technique (zoomed)

Wave variable technique with position feedforward component is also examined in this test. In the experiments, y-axis motion of the teleoperation is controlled by this new algorithm to enhance the position tracking performance. Figures 8.5-8.8 illustrate the results of this test.

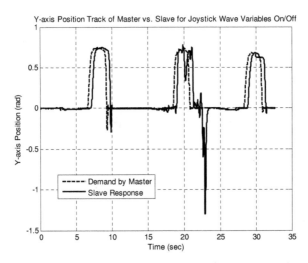

Figure 8.5. Position tracking performance of the identical master-slave teleoperation using the wave variable technique with position feedforward component

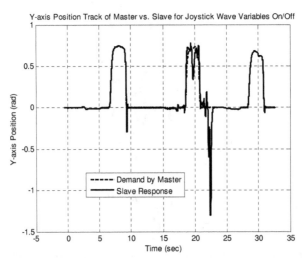

Figure 8.6. Position tracking performance of the identical master-slave teleoperation using the wave variable technique with position feedforward component (T_s-0.5 sec)

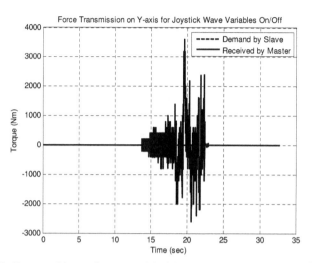

Figure 8.7. Force tracking performance of the identical master-slave teleoperation using the wave variable technique with position feedforward component

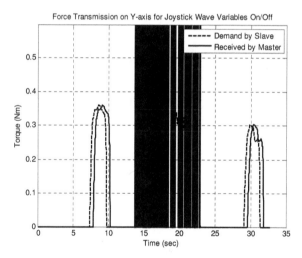

Figure 8.8. Force tracking performance of the identical master-slave teleoperation using the wave variable technique with position feedforward component (zoomed)

After the wave variable technique with position feedforward component is reactivated, it is observed from Figure 8.5 that not only the system is stabilized, but also the slave did not lose the track of the master's position trajectory. Figure 8.6 is created by shifting the slave position data 0.5 second back to eliminate the time delay effect in order to observe the position tracking performance clearly.

During the period when the wave variables are deactivated, the system becomes unstable and the slave reflects great amounts of forces. As the controller is switched on, it is clearly seen from Figure 8.8 that the master system did not lose track of the slave force demand.

The next test is conducted for the failure in which the communication line is deactivated for a limited period. During this time period, master and slave received null signals. The following results are for the x-axis, which is controlled by the customary wave variable technique.

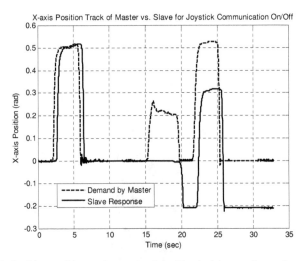

Figure 8.9. Position tracking performance of the identical master-slave teleoperation using the customary wave variable technique

The communication line is deactivated at the 15th second and reactivated at the 19th second. When the slave receives null signal as the velocity demand from the master, it does not react to the changes made in the master side. When the communication is established again, the slave tracks the position of the master stably but with an offset as observed in Figure 8.9. Although the position tracking is affected as a result of the communication line failure, force tracking performance does not show any drifts or instability as seen in Figure 8.10.

139

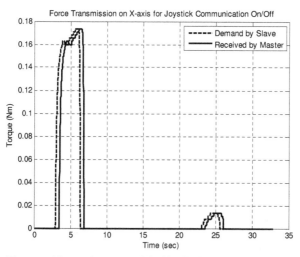

Figure 8.10. Force tracking performance of the identical master-slave teleoperation using the customary wave variable technique

In latter part of this experiment, the modified wave variable technique is examined on the y-axis. The results of this test are given in Figures 8.11–13.

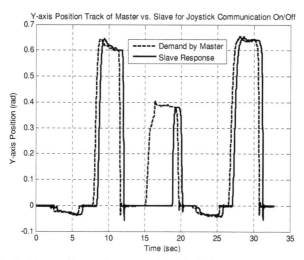

Figure 8.11. Position tracking performance of the identical master-slave teleoperation using the wave variable technique with position feedforward component

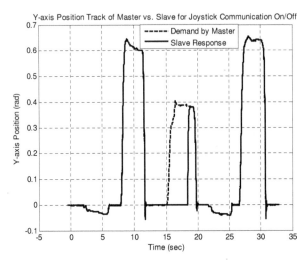

Figure 8.12. Position tracking performance of the identical master-slave teleoperation using the wave variable technique with position feedforward component (T_s-0.5 sec)

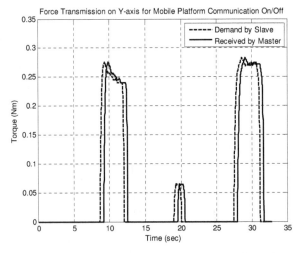

Figure 8.13. Force tracking performance of the identical master-slave teleoperation using the wave variable technique with position feedforward component

The y-axis of the teleoperation does not lose track of the position even after experiencing the communication loss as shown in Figure 8.11. In Figure 8.12, slave response data is shifted by 0.5 second to match the master position data in order to observe the tracking

141

performance better without the time delay. Force information sent from the slave is also tracked without any offsets or instability by the master except for the limited period when the communication line was down as shown in Figure 8.13.

8.2.2 Experimental Results with Variable Time Delays

The experimental setup used for the previous tests are also used for variable time-delay experiments. The position feedforward component is only used for the y-axis controller. The adaptive gain is applied for both axes as the system experiences variable time delays. The adaptive gain is switched off through the end of the experiments to observe the performance of the customary wave variable technique under variable time delays. The results of these experiments are also presented in [118]. Figure 8.14 shows the time delays during the experiments and the change in the gain as it adapts itself to the variations in time delays. The variable time delay profile is selected so that it is consistent with time delays over the Internet measured from the communication between Atlanta and Metz, France [24].

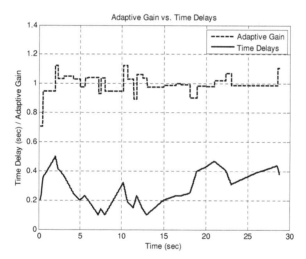

Figure 8.14. Changes in adaptive gain due to the time delays in identical master-slave teleoperation

The first set of figures represents the performance of the controller for the x-axis. The controller for this axis is the customary wave variables with the adaptive gain.

142

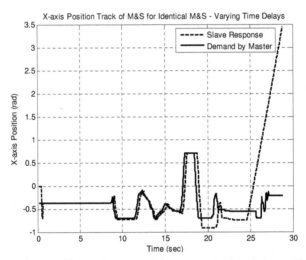

Figure 8.15. Position tracking performance of the wave variable technique with adaptive gain under variable time delays (*x*-axis)

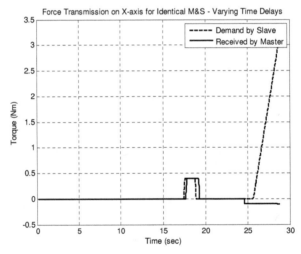

Figure 8.16. Force tracking performance of the wave variable technique with adaptive gain under variable time delays (*x*-axis)

The adaptive gain is switched off after the 24th second in the experiments. As it is observed from Figures 8.15 and 8.16, the system becomes unstable and the master cannot control the motions of the slave. It is also observed from Figure 8.15 that the slave lost the

143

position tracking of the master. Although the system does not experience any communication losses, a drift between the positions of the master and the slave is formed as a result of the variable time delays. No offset is observed in force tracking performance. Therefore, position feedforward component is employed for the controller on the y-axis. The following figures illustrate the results of the variable time delay experiments on the y-axis motion of the joystick.

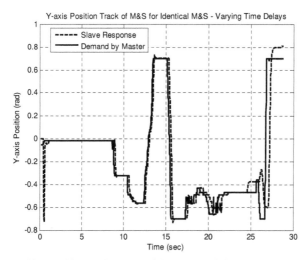

Figure 8.17. Position tracking performance of the wave variable technique with adaptive gain and position feedforward components under variable time delays (y-axis)

The slave does not lose track of the position when the position feedforward term is added to the controller with adaptive gain. This is clearly observed in Figures 8.17 and 8.18. The slave position does not even diverge from the demand after the adaptive gain is switched off after the 24th second. Instead, an offset is observed in Figure 8.19. Although the motion of the slave is kept stable with this controller without the adaptive gain, the force tracking is not possible and the system becomes unstable.

These experiments indicate that the wave variable technique with adaptive gain and position feedforward component is the most suitable controller for identical master and slave teleoperation experiencing variable time delays. Tracking of position and force demands is maintained and the system has remained stable throughout the experiments with this controller.

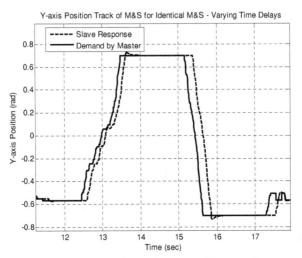

Figure 8.18. Position tracking performance of the controller on *y*-axis between the 11th and 18th seconds

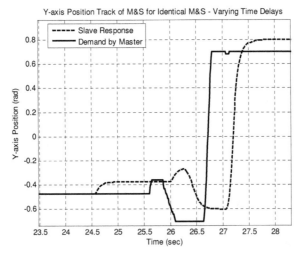

Figure 8.19. Position tracking performance of the controller on *y*-axis between the 23rd and 29th seconds

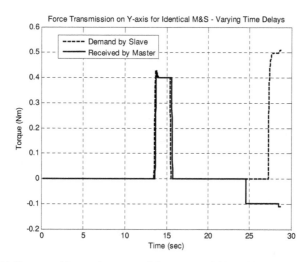

Figure 8.20. Force tracking performance of the wave variable technique with adaptive gain and position feedforward components under variable time delays (y-axis)

8.3. Redundant Slave Experiments

Two different experimental setups are used for redundant slave experiments. The tasks for both setups are so described that the slave industrial arms become redundant. The first setup is composed of the two-DOF gimbal-based master joystick and the Fanuc industrial arm. This setup is examined in position tracking performance tests. Phantom Omni Device and the Motoman industrial arm are used as the master and slaves systems for the variable time delay experiments. Both teleoperation systems also qualify as limited-workspace teleoperation since the slaves have predefined workspaces.

8.3.1 Experimental Results with Constant Time Delays

The Fanuc industrial arm model described in Chapter V is integrated to the teleoperation system as a virtual representation of the original manipulator. Gimbal-based joystick is also used for this teleoperation system as the master. This experimental setup is shown in Figure 8.21.

The teleoperation task is defined as tracing horizontal surfaces maintaining a point contact. While tracking the contour, the end-effector is required to maintain its orientation parallel to the normal of the surface. Therefore, only four DOF of the manipulator are used for the designed task. The last (fifth) joint is kept at a constant position throughout the tests.

146

First three joints are used for positioning while the fourth joint is used to maintain the orientation of the end-effector.

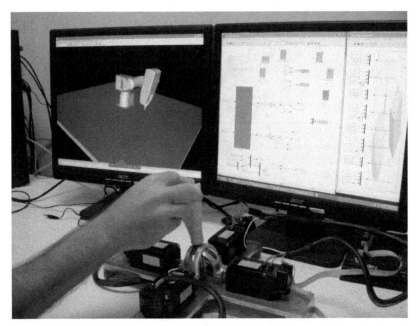

Figure 8.21. Experimental setup for gimbal-based master joystick and Fanuc slave arm

During the manipulation, if any of the joints two, three or four fails, the orientation objective can be sacrificed but the position tracking can be continued. This redundancy for the specific teleoperation task also promotes fault tolerance in the slave system.

The task for the experiments is defined as tracing a horizontal surface with obstacles. These obstacles enable creation of force information as the human operator runs into them operating the slave. The presence of these obstacles is observed by range sensors instead of force sensors. The readings from the range sensors are then transformed into force-reflection data. The tests are performed at 100 Hz sampling rate in real-time clock. The time delays for all of the tests are set at 0.5 second.

Customary wave variable technique is utilized for the first set of experiments. The second set of experiments employed the modified wave variable technique described in this paper. For both sets of tests, communication loss condition is considered. The results of these experiments are also presented in [119].

147

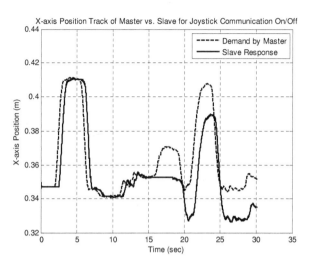

Figure 8.22. Position tracking performance on *x*-axis with customary wave variable technique
(Communication loss occurs between t=15 and 18 sec)

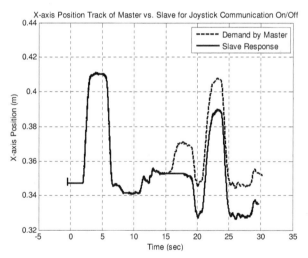

Figure 8.23. Position tracking performance on *x*-axis with customary wave variable technique
(T_s-0.5 sec)

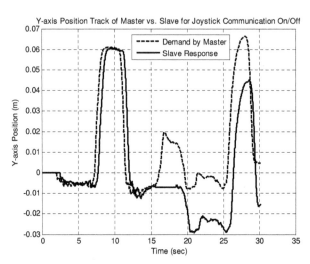

Figure 8.24. Position tracking performance on y-axis with the customary wave variable technique

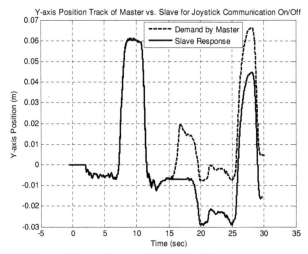

Figure 8.25. Position tracking performance on y-axis with customary wave variable technique (T_s-0.5 sec)

Figures 8.22 to 8.25 show the position tracking performance of the customary wave variable control obtained in the first set of experiments. Figures 8.23 and 8.25 are developed by plotting the slave position data 0.5 second prior to its occurrence to have a clear graph of

149

master-slave position tracking performance. The communication loss is realized between the 15th and 18th seconds of the telemanipulation task. During the communication loss, null data is received at the slave side, which makes the manipulator remain in the last position before the failure. As the communication is reestablished, the slave starts receiving velocity commands from the master but loses its position relative to the master as observed in the figures. The next set of graphs is developed for the experiments conducted by using the modified wave variable technique.

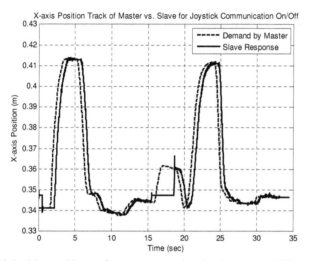

Figure 8.26. Position tracking performance on *x*-axis using the wave variable technique with position feedforward component (Communication loss t=16 to18 sec)

Figures 8.26-8.29 clearly indicate that the addition of a feedforward position demand compensates for steady state errors, which can be named as position drifts. At t=16 seconds, the communication is switched off and on again at t=18 seconds. During the time of no communication, the slave manipulator remains in its current position. As the slave starts receiving signals from the master, the steady state error is compensated and the position tracking resumes stably. The next set of figures shows the force translation from slave to the master. It is observed from Figure 8.30 and 8.31 that there are no problems in force tracking performance before or after the communication loss phenomenon.

150

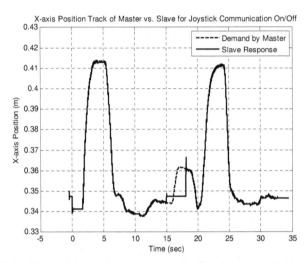

Figure 8.27. Position tracking performance on x-axis using the wave variable technique with position feedforward component (T_s-0.5 sec)

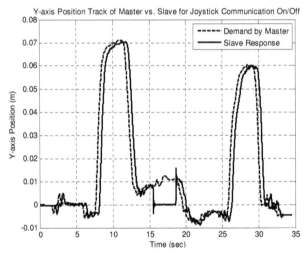

Figure 8.28. Position tracking performance on y-axis using the wave variable technique with position feedforward component

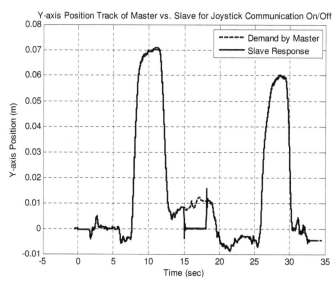

Figure 8.29. Position tracking performance on y-axis using the wave variable technique with position feedforward component (T_s-0.5 sec)

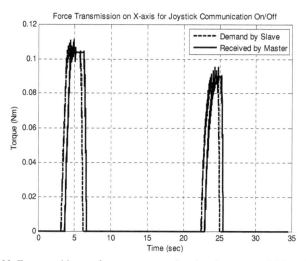

Figure 8.30. Force tracking performance on x-axis using the wave variable technique with position feedforward component

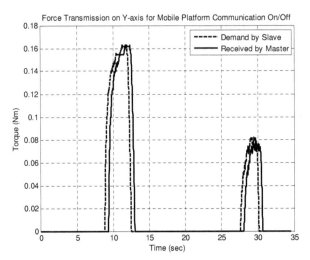

Figure 8.31. Force tracking performance on y-axis using the wave variable technique with position feedforward component

8.3.2 Experimental Results with Variable Time Delays

Two experimental studies are conducted with different redundant teleoperation systems for variable time delays. The first teleoperation system is composed of the Phantom Omni Device and the Motoman industrial arm. Phantom Omni Device is used as the actual master system. The virtual model of the Motoman UPJ manipulator is integrated to the system as the slave system. The model of the Motoman UPJ arm was described in Chapter V.

The commercial joystick described in Chapter IV and Phantom Omni Device is used to configure the second teleoperation system for these experiments. Phantom Omni Device is used as slave system in this experiment. The difference of this experiment from the first one is that actual devices are used for both the master and the slave.

8.3.2.1 Case Study 1

In the first experiment setup, the Motoman arm follows the position demands from the master. The orientation of the slave is to be maintained vertical to the horizontal ground while following the demands from the master. Because of this, only the joints one, two, three and five are used to position and orient the manipulator for this teleoperation setting. If any of the joints two, three or five fails, orientation requirements is degraded in order to continue the

153

task. Therefore, fault tolerance is achieved through the task definition for the slave manipulator. The setup for this experiment is presented in Figure 8.32.

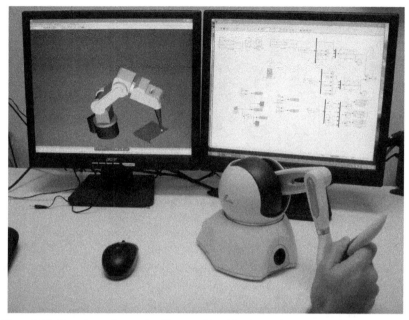

Figure 8.32. Experimental setup for Phantom Omni Device as the master and Motoman UPJ arm as the slave

In this teleoperation experiment, as the manipulator interacts with the ground, forces along the z-axis are fed to the actuators of the master. Also the friction forces along the x- and y-axes are fed to the master. The tests are performed at 1 kHz sampling rate. The variable time delays and the adaptive gain variation due to the changes in time delays used in this experiment are shown in Figure 8.33.

The experiment starts with system experiencing variable time delays. First, the customary wave variable technique is examined for this system and as the system initialized, the system became very instability and the virtual manipulator failed eventually. Therefore, the wave variable technique with the adaptive gain and position feedforward components is used initially. In this experiment, after a certain time of telemanipulation with the wave variable technique with adaptive gain and position feedforward component, the adaptive gain is removed to examine the necessity of the adaptive gain for variable time-delayed teleoperation. The following figures demonstrate the results of this experiment.

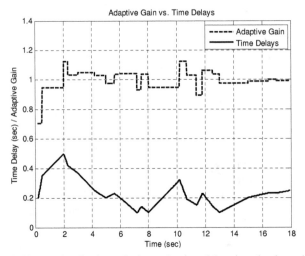

Figure 8.33. Changes in adaptive gain due to the time delays in redundant teleoperation

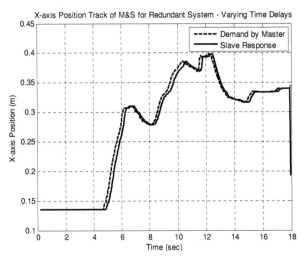

Figure 8.34. Position tracking performance on x-axis using the wave variable technique with position feedforward and adaptive gain on/off

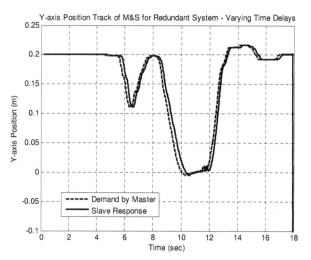

Figure 8.35. Position tracking performance on y-axis using the wave variable technique with position feedforward and adaptive gain on/off

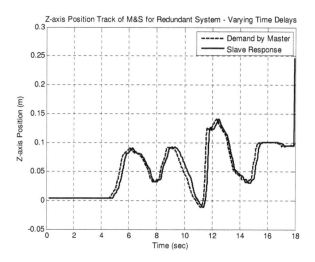

Figure 8.36. Position tracking performance on z-axis using the wave variable technique with position feedforward and adaptive gain on/off

It is observed from Figures 8.34 to 8.36, the wave variable technique with adaptive gain and position feedforward component has a satisfactory behavior under variable time delays. The adaptive gain component is switched off just before the 18th second in this

156

experiment. The slave diverges from the demand as soon as this gain is switched off and the system fails due to instabilities. The next set of figures illustrates the force tracking performance of this new controller.

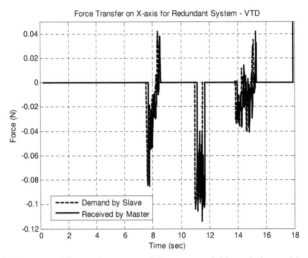

Figure 8.37. Force tracking performance of the wave variable technique with adaptive gain and position feedforward components under variable time delays (x-axis)

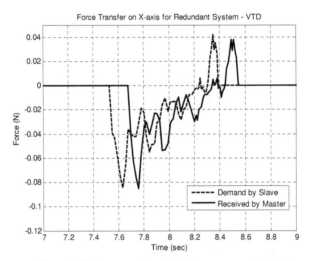

Figure 8.38. Force tracking performance zoomed (x-axis)

157

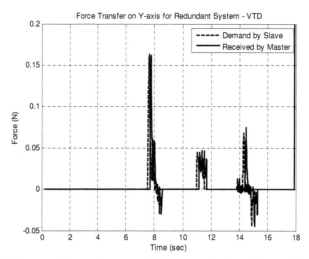

Figure 8.39. Force tracking performance of the wave variable technique with adaptive gain and position feedforward components under variable time delays (*y*-axis)

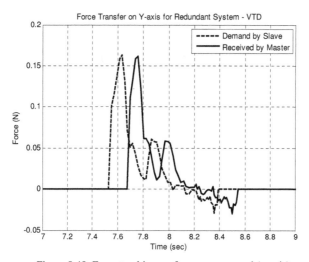

Figure 8.40. Force tracking performance zoomed (*y*-axis)

The force applied along the normal of the surface (z-axis) is presented in Figure 8.41. Figures 8.37 to 8.39 illustrate the forces created due to the surface friction. Figures 8.38, 8.40 and 8.42 provide a detailed view of force tracking performance along the three axes. It is observed from the force tracking figures that until the 18th second, the system is stable and forces are tracked without any drifts. The adaptive gain is switched off just before the 18th second, which causes the system to become instable and apply excessive amounts of forces.

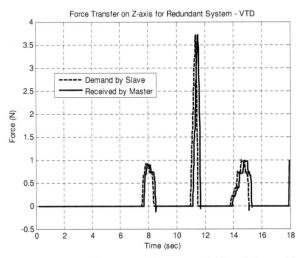

Figure 8.41. Force tracking performance of the wave variable technique with adaptive gain and position feedforward components under variable time delays (z-axis)

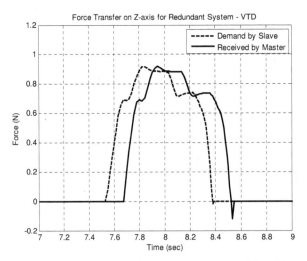

Figure 8.42. Force tracking performance zoomed (z-axis)

The results of this experiment indicate that the wave variable with adaptive gain and position feedforward component is necessary to keep the system stable under variable time delays. Even if one of these components is inactive, complex systems that involve telemanipulation of serial arms cannot continue teleoperating and fail eventually.

8.3.2.2 Case Study 2

The commercial force-feedback joystick presented in Chapter IV is used as the master system and the Phantom Omni Device is used as the slave system in this teleoperation experiment. The operator using the master system drives the slave in two axes. On the other side of the teleoperation, another operator holds the stylus of the Phantom Omni Device. Slave side operator is forced to follow the demands from the master. As the slave side operator resists following the demands, this resistance is recognized as the environmental forces and they are reflected to the master system. The rule to recognize the resistance is declared as feeding in a force more than 0.3 N to the Phantom Omni Device in any direction. The experimental setup is illustrated in Figure 8.43.

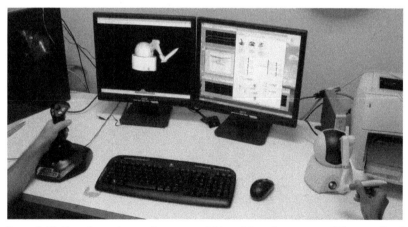

Figure 8.43. Experimental setup for commercial joystick as the master and Phantom Omni Device as the slave

The experiments are performed at 1 kHz sampling rate. The variable time delays and the adaptive gain variation due to the changes in time delays used in this experiment are shown in Figure 8.44.

160

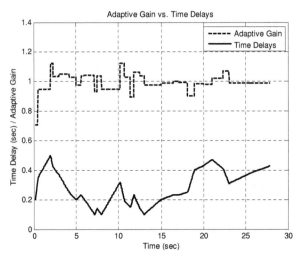

Figure 8.44. Changes in adaptive gain due to the time delays in redundant teleoperation with the Logitech joystick and the Phantom Omni Device

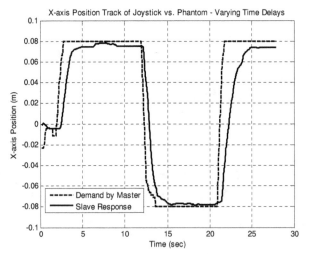

Figure 8.45. Position tracking performance on x-axis using the wave variable technique with position feedforward and adaptive gain components

The wave variable technique with position feedforward and adaptive gain is always the active controller throughout this experiment. The main reason for this is to protect the actual devices from excessive amounts of forces as the system is shown to be unstable with

161

the other controllers in the previous experiment. The following figures present the position tracking performance of the master and the slave in this experiment.

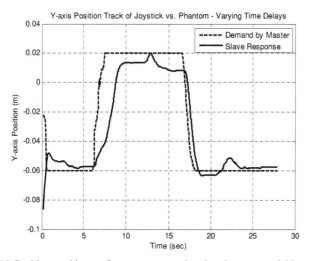

Figure 8.46. Position tracking performance on *y*-axis using the wave variable technique with position feedforward and adaptive gain components

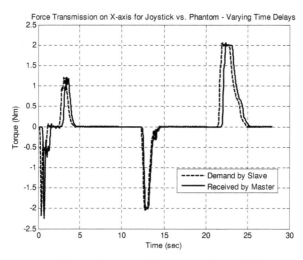

Figure 8.47. Force tracking performance of the wave variable technique with adaptive gain and position feedforward components under variable time delays (*x*-axis)

It is observed from Figures 8.45 and 8.46 that the system is stable but there are position errors in both axes. The reason for these errors is that the slave side is constantly under the control of the slave side operator. The operator resists to the motion demands as they are issued. The effects of this resistance are shown in Figures 8.47 and 8.48 as the slave forces reflected back to the master. These figures provide a better idea on the position error observed in Figures 8. 45 and 8.46.

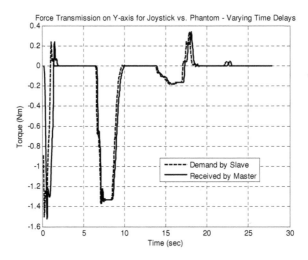

Figure 8.48. Force tracking performance of the wave variable technique with adaptive gain and position feedforward components under variable time delays (y-axis)

It is clearly observed from Figure 8.47 that the slave operator resists to the motion demands between the 22nd and 25th seconds. The slave forces increase as the slave system is forced not to follow the motion demands as observed in Figure 8.45 between the 22nd and 25th seconds. Similar behavior is observed in the motions along the y-axis. The slave side operator resists to the motion demands and slave forces increase between the 6th and 10th seconds. As a result of this, slave cannot follow the motion demands from the master precisely during this period. The error in position tracking along y-axis can be observed between 6th and 10th seconds in Figure 8.46.

The results of this experiment show that the teleoperation system composed of actual robotic devices can be stabilized under time delays using the wave variable technique with position feedforward and adaptive gain components. The tracking performances are also

163

shown to be satisfactory for both position and force demands. A low-pass filter is used to filter out high frequency oscillations observed in force calculations to smoothen the telemanipulation.

8.4. Position/Force Controller Experiments

The slave remote system selected for these experiments is the four-DOF SCARA manipulator, Epson E2H853C introduced in Chapter V. Therefore, it has a limited workspace and requires mapping between the joystick motion and the slave motion. The master system is the gimbal-based joystick. The experimental setup for this study is given in Figure 8.49. The results of this experiment are also presented in [120].

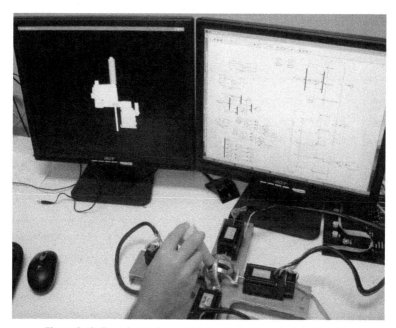

Figure 8.49. Experimental setup for the gimbal-based master joystick
and Epson SCARA

The teleoperation task is designed to trace surfaces by exerting controlled amounts of forces. The commands received in the x- and y-axes are transmitted to the slave as Cartesian coordinate inputs. Then the forces created during the telemanipulation as a result of interaction and surface friction are fed back to the master.

Interaction model between the end-effector and the surface is created using a planar contact. The material of the end-effector is selected to be lead with a modulus of elasticity of 36.5 GPa. The surface to be traced is assumed to have a relatively high stiffness value compared to the robot; therefore, it is modeled as rigid. The friction between the end-effector and the surface is created using Coulomb friction and stiction model. The coefficient of friction used in the experiment is of lead which is specified as 0.43 and stiction is regulated at 1.00.

The operator specifies the desired force to be applied to the environment through an input screen. This force profile can also be called the force trajectory. The interaction force between the surface and the end-effector as well as the surface friction force are created in the simulation environment. The friction forces in Cartesian space are then fed back into the servomotors of the joystick's respective axes. The friction forces are also sent back to the master as force feedback information. The tests are run in 1 kHz sampling rate with the configuration described above.

In the simulations, independent joint control is used. This provides the flexibility of using different control algorithms for each joint. In the first set of simulations, Proportional-Derivative (PD) control was implemented for every joint. After tuning the PD parameters, the controller produced an acceptable error range when there was no contact with the surface. In order to control the force applied on the surface with this controller, the desired position trajectory was modified to touch the surface so as to create the desired amount of contact force.

As expected, a pure position controller for the prismatic joint was not effective enough to follow the force trajectory while following the position trajectory. Another solution is to use dual controls. A position controller makes the end-effector approach the surface and create the contact. Then the control algorithm has to be switched to a pure force controller or to hybrid position/force controller to follow the force trajectory. This type of switching between the controls can cause instabilities and chattering in a teleoperation system where the communication can be delayed by an unacceptable period. One other possibility for a system that is required to follow both position and force trajectories is to use the modified admittance controller presented in Chapter VI.

The following set of experiments is carried out by using a modified admittance controller for the prismatic joint and PD controller for the revolute joints. The PD control parameters were modified to compensate for the disturbances created by the friction forces at revolute joints.

165

The teleoperation task is defined as to follow the Cartesian coordinate demands by the master joystick while following the force trajectory specified by the operator. The operator starts the telemanipulation at null position and then makes the contact with the surface by applying force on it. Later, the operator follows a square path inside the workspace of the slave. While executing the task, friction forces are also calculated as fed back into the joints as disturbance as well as to the actuators of the master systems. As the actuators of the master system are driven with these force feedbacks, operator gets the feeling that he/she is in contact with the surface and friction forces are acting on the opposite direction of the motion.

The trajectory of the end-effector is shown in the below four graphs in x-y plane, along x-axis and y-axis, and then along z-axis respectively. The x-y graph indicates that the manipulation started at the null position and then a 0.4 by 0.4 meter square path is followed by the slave as demanded by the operator through the master joystick.

It is observed from Figures 8.51 and 8.52 that the operator starts to follow a square path at the 15th second and finished by the 23rd second of the telemanipulation. Slave manipulator follows the demands from the master joystick with minimal transition periods as set by the PD control gains.

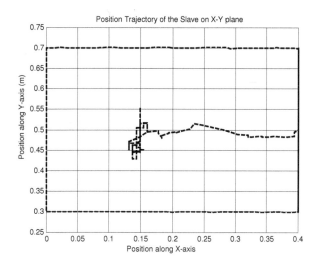

Figure 8.50. End-effector trajectory in x-y plane

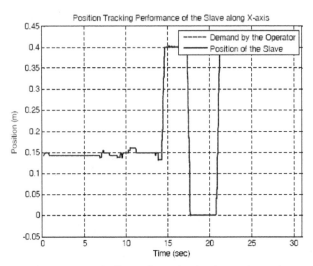

Figure 8.51. End-effector demand and trajectory in *x*-axis

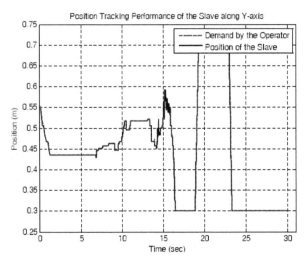

Figure 8.52. End-effector demand and trajectory in *y*-axis

Figure 8.53 shows the trajectory of the end-effector along z-axis. The end-effector approaches the surface at the 9th second and as the force trajectory is specified as larger than no-force, the end-effector moves towards the surface and makes contact to apply the demanded force.

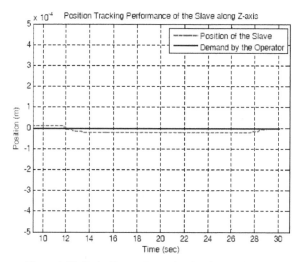

Figure 8.53. End-effector demand and trajectory in z-axis

The force applied to the surface is presented in Figure 8.54. It is observed that after an acceptable transition period, the force trajectory demand is followed without any overshoots. The transition state characteristics can be changed by modifying the admittance term of the controller. The position control law however is kept constant as a PD controller. It was not necessary to modify the position control parameters. The position was tracked in acceptable error range with the inner position loop of the admittance controller. It is also seen that at the time of the first approach to the surface, the end-effector hits the surface creating about 0.5 N of force for a limited period. This is due to the overshoot at the position controller. At the time of the first contact, the force demand is zero; therefore, the end-effector is pulled back not to apply any forces by the outer force controller loop.

The tests show that the virtual slave manipulator follows the force trajectory within an acceptable error range and with acceptable transients while following the motion trajectory. It also creates the force feedback information in terms of friction along the x- and y-axis, as it is observed in Figures 8.55 and 8.56 for the operator to provide a feel of the remote environment.

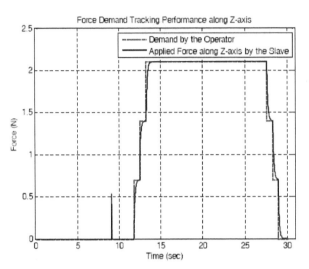

Figure 8.54. Force applied by the end-effector to the surface

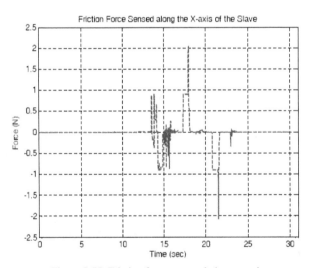

Figure 8.55. Friction forces created along x-axis

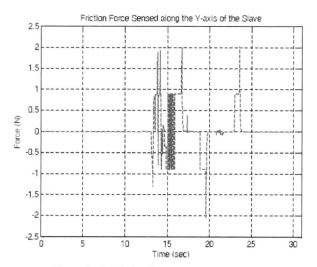

Figure 8.56. Friction forces created along y-axis

8.5. Conclusions

All teleoperation systems require a communications line to connect the master and the slave. In most cases, there are large enough time delays in the communications line that cause instabilities in the system. Additionally, communications lines are often a major source of failure and interrupt communication between the two subsystems. This failure in limited periods causes drifts in position tracking.

In limited-workspace teleoperation, experiments indicate that the customary wave variable control fails to eliminate this drift even though it eliminates instability due to time delays. A position feedforward component is proposed and the experimental results indicate that this proposed modification corrects the drift successfully.

Position tracking is the first priority in limited-workspace teleoperation systems. Therefore, the wave variable technique with the position feedforward component created satisfactory results in limited-workspace, constant time-delayed teleoperation systems.

It is also noted that the force information sent from the slave is tracked without any drifts by the master in limited-workspace teleoperation systems tested in this chapter. Therefore, it is concluded that the usage of a force feedforward component for the master side is unnecessary.

The control algorithms examined for limited-workspace teleoperation systems experiencing constant time delays cannot guarantee stability under variable time delays. An adaptive gain component is introduced in Chapter VI to be used in such cases. As a result of the addition of this component, limited-workspace teleoperation systems under variable time delays are stabilized. The experiments have also shown that the system eventually becomes unstable when the adaptive gain becomes inactive. It is also concluded that the position feedforward component is necessary to keep the system stable and to track the position demands under variable time delays.

The wave variable technique with adaptive gain and position feedforward components is also used to control an actual master-slave limited-workspace teleoperation system. The results of the experiments with this system also proved that the proposed controller can stabilize the system under variable time delays while tracking the position and force demands successfully.

Also, in this chapter, two widely used control algorithms are examined and modified for use in limited-workspace teleoperation. The main objective was to compare the control algorithms for stability in case of time delays or communication losses.

The customary versions of the hybrid position/force and admittance controllers fail in teleoperation system applications. The reason for the failure is the assumption that the error between the demanded and the actual positions in Cartesian space is small for a finitely small amount of time. This assumption becomes invalid when the system experiences time delays or loss of communication.

The assumption used in customary controls is not necessary for modified versions of these controllers when the comparison between the demanded and the actual signals is made in joint space. On the other hand, hybrid position/force control requires a pure position controller for switching from one to another while in transition from contact to no contact. Therefore, possible instabilities and chattering are foreseen in case of time delays or communication losses.

The modified admittance controller does not require switching to a pure position controller regardless of the contact condition. The outer force controller loop is always active and ready for force demands. The experimental results indicate that among the controllers examined in this chapter, admittance controller is seen as the best choice for teleoperation applications that may experience communications line problems.

CHAPTER IX

UNLIMITED-WORKSPACE TELEOPERATION EXPERIMENTS

9.1. Introduction

Teleoperation systems composed of unlimited-workspace slave systems such as mobile platforms, aerial vehicles or submarines are categorized as unlimited-workspace teleoperation. In this chapter, experimental studies are continued for the unlimited-workspace teleoperation systems. The unlimited-workspace slave used in the experiments is the holonomic mobile platform presented in Chapter V.

The holonomic mobile platform has three uncoupled DOF. Out of these three, only two of them are used in the tests. This slave is configured as a fault-tolerant system since it is powered by four omni-directional wheels. When a wheel fails, the remaining three wheels are capable to compensate for the failed wheel and continue executing the demands sent from the master.

These motions are along the x- and y- axes. Therefore, the position changes detected in x- and y-axes of the master joystick are mapped to the velocity changes of the mobile platform along these axes. This type of mapping enables the mobile platform to have unlimited workspace even though it is controlled with a limited-workspace master joystick. The gimbal-based master joystick and the virtual model of the holonomic mobile platform are used in the experiments with both constant and variable time delays. The commercial joystick from Logitech and the holonomic mobile platform constructed in this work are also used in experiments with variable time delays.

9.2. Experimental Results with Constant Time Delays

In the constant time delay experiments, initially, the customary wave variable technique is used for both axes. The time delay is kept constant at 0.5 second. Two sets of experiments are presented in this section. The first set of experiments is conducted while the wave variable technique is deactivated for a limited time. Communication line is switched off-and-on during the second set of experiments. In these experiments, gimbal-based joystick is used as the master system and the virtual model of the holonomic mobile platform is used as the slave system as shown in Figure 9.1. The results of both sets of experiments are also presented in [117].

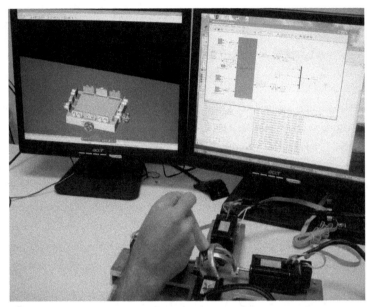

Figure 9.1. Experimental setup for gimbal-based master joystick and the virtual model of the holonomic mobile platform

9.2.1 Experiments with an Unstable System

The following graphs are created for a teleoperation system experiencing constant time delays. Initially the system is controlled with the customary wave variable technique. Towards the middle of the experiment, the customary wave variable technique is switched off and the demands are fed directly to each subsystem without any control over them. This introduced instability to the system. The wave variables are then reactivated to stabilize an unstable system under constant time delays.

The main objective of the unlimited-workspace teleoperation is to control the velocity of the slave with the position commands from the limited-workspace master. Therefore, the tracking priority of this type of teleoperation is the velocity. The customary wave variable technique provides acceptable results for this goal. There are no instabilities or drifts in the velocity tracking even when the system is unstable prior to the reactivation of the wave variable technique.

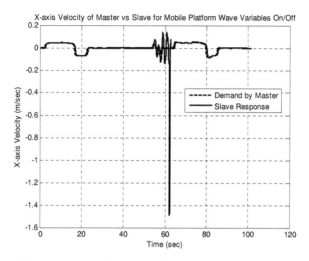

Figure 9.2. Velocity tracking performance of the unlimited-workspace teleoperation using the customary wave variable technique – wave variables on/off (*x*-axis)

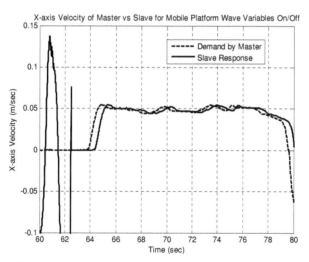

Figure 9.3. Velocity tracking performance along *x*-axis – wave variables on/off (zoomed)

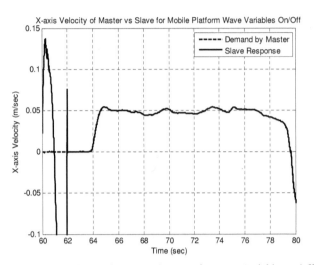

Figure 9.4. Velocity tracking performance along x-axis – wave variables on/off (zoomed & T_s-0.5 sec)

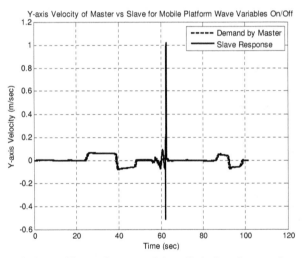

Figure 9.5. Velocity tracking performance of the unlimited-workspace teleoperation using the customary wave variable technique – wave variables on/off (y-axis)

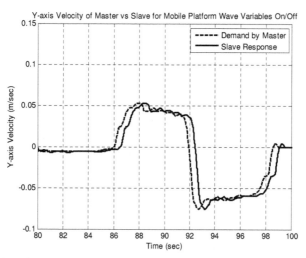

Figure 9.6. Velocity tracking performance along y-axis – wave variables on/off (zoomed)

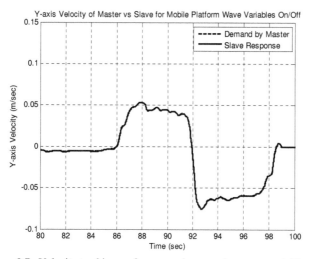

Figure 9.7. Velocity tracking performance along y-axis – wave variables on/off
(zoomed & T_s-0.5 sec)

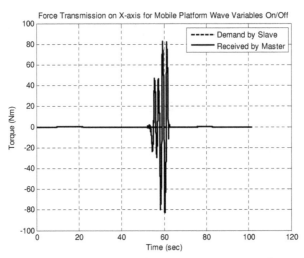

Figure 9.8. Force tracking performance of the unlimited-workspace teleoperation using the customary wave variable technique – wave variables on/off (x-axis)

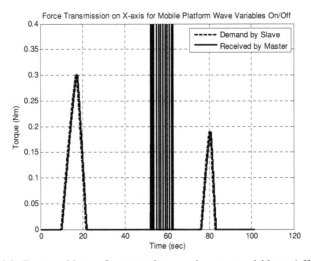

Figure 9.9. Force tracking performance along x-axis – wave variables on/off (zoomed)

177

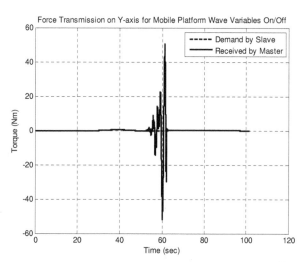

Figure 9.10. Force tracking performance of the unlimited-workspace teleoperation using the customary wave variable technique – wave variables on/off (*y*-axis)

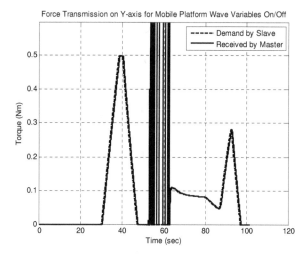

Figure 9.11. Force tracking performance along *y*-axis – wave variables on/off (zoomed)

9.2.2 Experiments with Communication Failure

The effect of communication loss for limited periods is examined in this subsection. The communications line is deactivated during the experiment for a limited period and then reactivated to examine the tracking performance and the stability of the customary wave variable technique under constant time delays. During the communication loss, a null signal is transmitted to both ends of the teleoperation system. The slave system received a zero velocity demand, while the master received a zero force reflection from the slave side. Therefore, the velocity and force outputs of the slave and the master are pulled to zero during this period. This period of communication loss lasted for a few seconds and then the communication is reestablished. The tracking performance is then observed for the position and force demands.

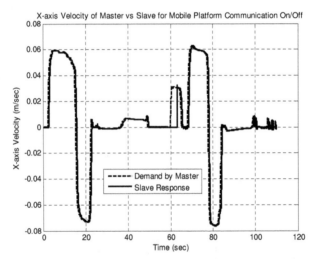

Figure 9.12. Velocity tracking performance of the unlimited-workspace teleoperation using the customary wave variable technique – communications on/off (x-axis)

Both of the teleoperation subsystems received null signals during the communication loss. As the communication between the two subsystems is established, the slave did not lose track of the velocity along the x- and the y-axes as seen in Figures 9.12 and 9.15. In Figures 9.13 and 9.16, the establishment of communication from the no-communication period is clearly observed. It is also observed that the limited period of communication loss has not

179

caused any drifts in velocity tracking performance of the unlimited-workspace teleoperation under constant time delays.

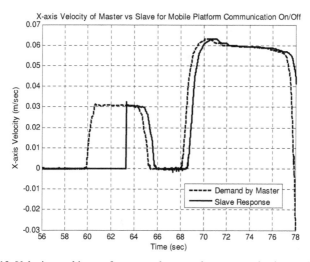

Figure 9.13. Velocity tracking performance along x-axis – communications on/off (zoomed)

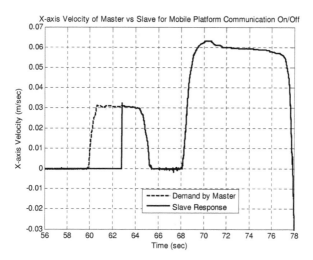

Figure 9.14. Velocity tracking performance along x-axis – communications on/off (zoomed & T_s-0.5 sec)

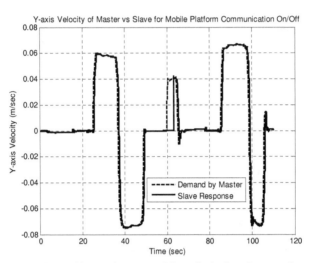

Figure 9.15. Velocity tracking performance of the unlimited-workspace teleoperation using the customary wave variable technique – communications on/off (y-axis)

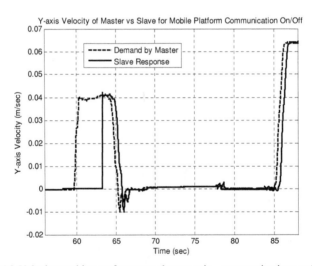

Figure 9.16. Velocity tracking performance along y-axis – communications on/off (zoomed)

181

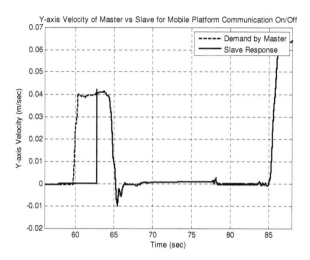

Figure 9.17. Velocity tracking performance along y-axis – communications on/off (zoomed & T_s-0.5 sec)

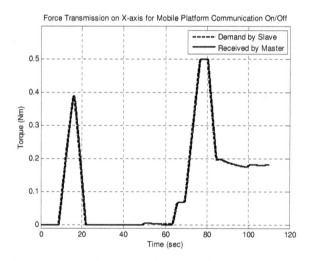

Figure 9.18. Force tracking performance of the unlimited-workspace teleoperation using the customary wave variable technique – communications on/off (x-axis)

182

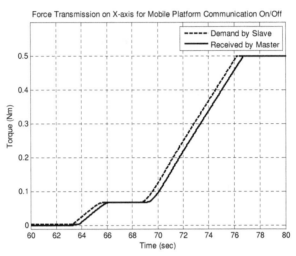

Figure 9.19. Force tracking performance along x-axis – communications on/off (zoomed)

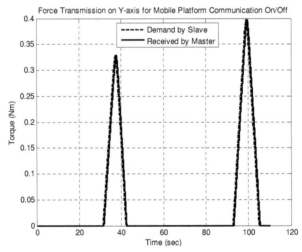

Figure 9.20. Force tracking performance of the unlimited-workspace teleoperation using the customary wave variable technique – communications on/off (y-axis)

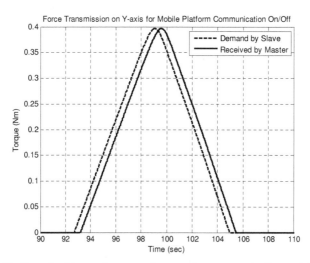

Figure 9.21. Force tracking performance along *y*-axis – communications on/off (zoomed)

As in the previous cases, the slave velocity data is shifted by 0.5 second in Figures 9.14 and 9.17 to match the master demand. Also, the master has not lost track of the force information sent by the slave in both axes. This is observed in Figures 9.18 and 9.20. Figures 9.19 and 9.21, on the other hand, present zoomed graphs of force tracking in both axes after the reactivation of the communication line. These figures also indicate that the force tracking is not affected by limited periods of communication losses.

9.3. Experimental Results with Variable Time Delays

Two experimental setups are used to examine unlimited-workspace teleoperation systems under variable time delays. The first system consists of the gimbal-based joystick as the master systems and the virtual model of the holonomic mobile platform. The second teleoperation configuration is composed of the commercial Logitech joystick and the actual holonomic mobile platform. The task for both systems is defined as navigating with the demands received from the master and reflecting forces as the slave approaches obstacles.

9.3.1 Case Study 1

Two-DOF gimbal-based joystick controls the virtual model of the mobile platform in this case study as shown in Figure 9.1. The variable time delays and the adaptive gains used for this experiment are presented in Figure 9.22. The variable time delay profile is selected so that it is consistent with time delays over the Internet measured from the communication between Atlanta and Metz, France [24].

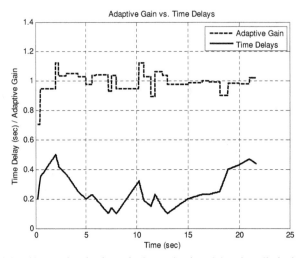

Figure 9.22. Changes in adaptive gain due to the time delays in unlimited-workspace teleoperation with virtual slave

The initial controller for the experiments is the wave variable technique with adaptive gain. This controller is later switched to the customary wave variable controller. Both axes of the holonomic mobile platform are controlled with the same controller. The following figures present the experimental results for the teleoperation of the holonomic mobile platform under variable time delays. During this experiment, the human operator drives the mobile platform towards the preset obstacles. As the platform moves closer to the obstacles, forces are reflected to restrict further motion along that direction. At the 18th second of the telemanipulation, the adaptive gain component is switched off and the teleoperation controller converts to the customary wave variable technique.

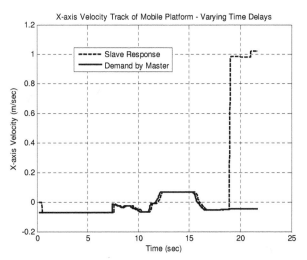

Figure 9.23. Velocity tracking performance of the unlimited-workspace teleoperation with virtual slave under variable time delays (*x*-axis)

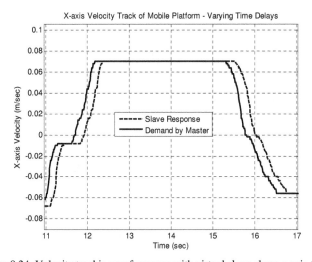

Figure 9.24. Velocity tracking performance with virtual slave along *x*-axis (zoomed)

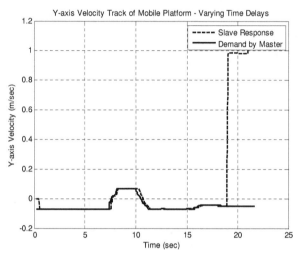

Figure 9.25. Velocity tracking performance of the unlimited-workspace teleoperation with virtual slave under variable time delays (*y*-axis)

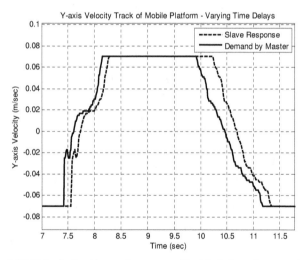

Figure 9.26. Velocity tracking performance with virtual slave along *y*-axis (zoomed)

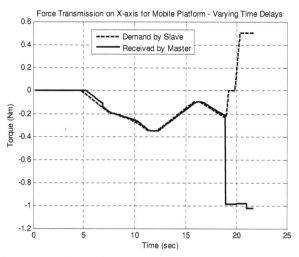

Figure 9.27. Force tracking performance of the unlimited-workspace teleoperation with virtual slave under variable time delays (x-axis)

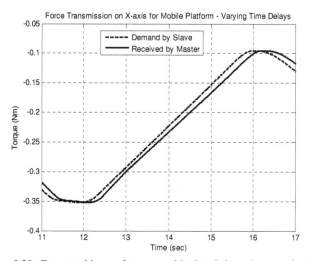

Figure 9.28. Force tracking performance with virtual slave along x-axis (zoomed)

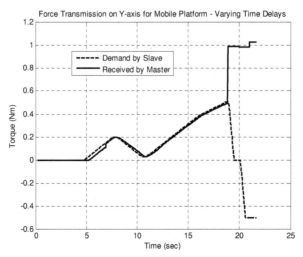

Figure 9.29. Force tracking performance of the unlimited-workspace teleoperation with virtual slave under variable time delays (*y*-axis)

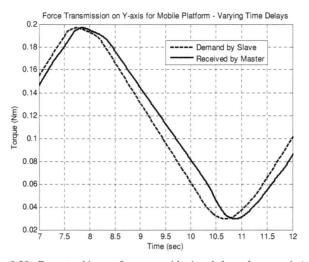

Figure 9.30. Force tracking performance with virtual slave along *y*-axis (zoomed)

As explained earlier, the tracking priority in unlimited-workspace teleoperation systems is usually the velocity. Therefore, it is clearly observed from Figures 9.23, 9.24, 9.25 and 9.26 that the velocity tracking performance of the wave variable technique with adaptive gain is satisfactory. The position feedforward component is not required to stabilize unlimited-workspace teleoperation systems under variable time delays. The force tracking performance is also satisfactory with the controller used in this experiment (Figures 9.28 and 9.30). After the 18th second of this experiment, the adaptive gain is switched off. This changed the controller to the customary wave variable technique. As a result of this, the system became unstable. This effect is clearly observed from the velocity tracking performances (Figures 9.23 and 9.25) and the force tracking performances along both axes (Figures 9.27 and 9.29).

9.3.2 Case Study 2

The commercial joystick described in Chapter IV and the actual holonomic mobile platform is used to configure the teleoperation system for this experiment as shown in Figure 9.31. The experiment is conducted in a sampling rate of 20 Hz. The main reason for this is the communication speed limitations through the serial port of the computer. Although the sampling rate is lower than the previous experiments, teleoperation system is still able to operate since the mobile platform travels in relatively slow speeds (V_{max} = 80 mm/sec). A source code working in the microprocessor of the mobile platform supplies the necessary outputs to drive the servomotors while obtaining the sensory information from the range sensors. An interface written in Matlab$^{©}$ M-file programming language is also created to send velocity demands to the microprocessor of the mobile platform and receive sensory information. Both source codes are given in Appendix C.

The wave variable technique with adaptive gain component is used to control both axes throughout the experiment. The change in adaptive gain due to the time delays is given in Figure 9.32.

The following figures show the mobile platform approaching to obstacles along the x- and y-axes. The first set of figures is drawn for the velocity tracking performance of the teleoperation system. The slave velocity shown in the figures is velocity demand received at the slave side from the communications line. The reasons for this are that the slave system does not have sensors to measure its velocity and the sonar sensor have limited range which makes them useless in measuring speeds when the system is not close to the obstacles.

Figure 9.31. Experimental setup for the gimbal-based master joystick and the holonomic mobile platform

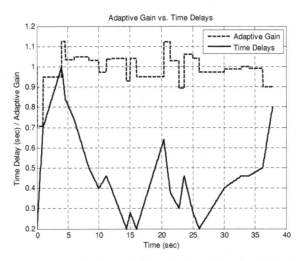

Figure 9.32. Changes in adaptive gain due to the time delays in unlimited-workspace teleoperation with actual slave

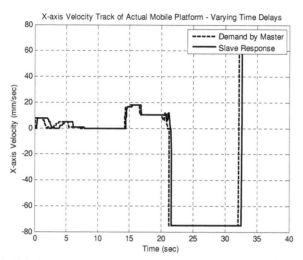

Figure 9.33. Velocity tracking performance of the unlimited-workspace teleoperation with actual slave under variable time delays (*x*-axis)

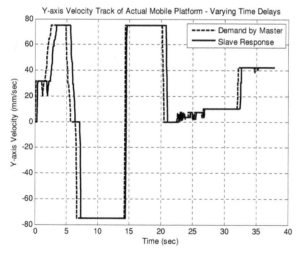

Figure 9.34. Velocity tracking performance of the unlimited-workspace teleoperation with actual slave under variable time delays (*y*-axis)

Figure 9.33 and 9.34 indicate that the system is stable and velocity tracking performance is satisfactory using the wave variable technique with the adaptive gain. The next set of figures shows the force tracking performance of this teleoperation system as the

192

slave approaches obstacles. The distance measured by using range sensors on the slave system is then converted to forces to be sent to the master. The forces are regulated as a result of this conversion and necessary amounts of torques are applied to the actuators of the joystick in order to make the human operator feel the presence of an obstacle.

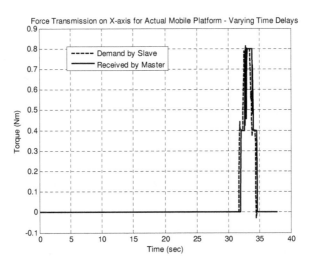

Figure 9.35. Force tracking performance of the unlimited-workspace teleoperation with actual slave under variable time delays (x-axis)

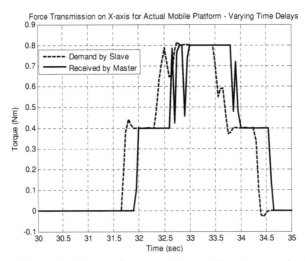

Figure 9.36. Force tracking performance with actual slave along x-axis (zoomed)

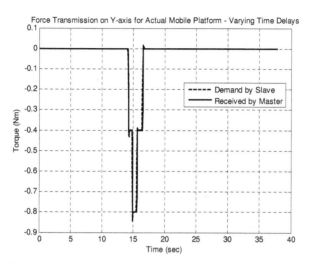

Figure 9.37. Force tracking performance of the unlimited-workspace teleoperation with actual slave under variable time delays (y-axis)

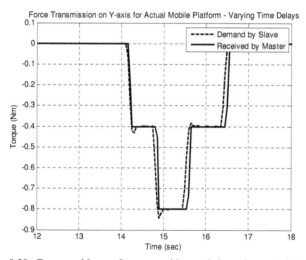

Figure 9.38. Force tracking performance with actual slave along y-axis (zoomed)

It is clearly observed from Figures 9.35 and 9.37 that the system is stable throughout the experiment and excessive amounts of forces are not formed as a result of any instability. Figures 9.36 and 9.38 indicate that the force tracking performance is satisfactory along both axes using the wave variable technique with adaptive gain.

9.4. Conclusions

In unlimited-workspace teleoperation systems, the goal is generally to track the velocity commands issued by the master. In light of the experimental results, it is concluded that the customary wave variable technique is sufficient for the slave to track master's velocity commands and the master to track slave's force demand without any drifts or instabilities. Therefore, the customary wave variable technique created satisfactory results for unlimited-workspace, constant time-delayed teleoperation systems.

The experimental results with constant time delays also indicate that the force information sent from the slave is tracked without any drifts by the master using the customary wave variable technique. Therefore, it is concluded that the usage of a force feedforward component for the master side is unnecessary.

The control algorithms examined for unlimited-workspace teleoperation systems experiencing constant time delays cannot guarantee the system stability under the variable time delays. An adaptive gain component is introduced in Chapter VI for such cases. As a result of the addition of this component, unlimited-workspace teleoperation systems under variable time delays are stabilized. In the experiments with the virtual slave, it is also shown that when the adaptive gain becomes inactive the system eventually becomes unstable.

The wave variable technique with adaptive gain is also employed to control an actual master-slave unlimited-workspace teleoperation system. Although the sampling rate is limited to 20 Hz for this experiment, the proposed controller produced satisfactory results in maintaining the stability as well as tracking the velocity and force demands.

CHAPTER X

CONCLUSIONS

10.1. Conclusions and Findings

In this work, teleoperation is first defined as an area of robotics where two robotic systems cooperate to accomplish a task. The components of a teleoperation system are listed as the master system, slave system and the communications line that connects these two systems. The master is controlled by the human operator to send motion commands to the slave. The slave is then driven by these commands while interacting with its environment. This task usually takes place at a hazardous environment or at a distant site from the operator. In both cases, humans cannot carry out the mission and a robotic system is required. As the slave interacts with the environment, it sends sensory information back to the master system enabling the operator to feel the slave-environment interaction. This is called telepresence. The level of telepresence can be increased by sending back more sensory information. The sensory information used in this study is the force and the visual feedbacks.

In Chapter II, previous teleoperation applications are overviewed. Later, teleoperation systems are categorized as unilateral or bilateral depending on the nature of information flow. Unilateral teleoperation only sends motion demands to drive the slave where the slave does not provide any feedback signals. In bilateral teleoperation, as the master sends motion demands to the slave, the slave sends back sensory feedbacks. The special type of bilateral teleoperation that uses force feedback signals is titled force-reflecting teleoperation. In this work, this type of teleoperation is classified as either limited- or unlimited-workspace teleoperation depending on the nature of the slave's workspace. Following this, fault tolerance concept and its implementation in different levels are introduced as they apply to teleoperation systems. Finally, teleoperation system architecture is formulated in a flowchart format to guide the design engineer develop the conceptual design of a teleoperation system more efficiently.

A new system modeling method developed as part of this work, Virtual Rapid Robot Prototyping (VRRP) as well as other modeling methods are described in Chapter III. The new method is shown to save time while modeling the systems accurately. In order to achieve this, critical steps in Computer-Aided-Design (CAD) modeling and also mechanical system modeling environment, Matlab$^{©}$ Simmechanics are reviewed. The system models created with this method are successfully used in the simulation studies.

In Chapter IV, slave system models created by using VRRP are used to configure virtual haptic laboratory environments. The integration of actual controller devices (master systems) to manipulate virtual slaves is explained. This developed virtual haptic environment does not only lend itself to experimental setups for testing and verification of controllers, but also is an efficient and inexpensive training simulator. This tool is general purpose and capable of creating almost any environment that can be modeled in CAD and integrating any controller device with a PC interface.

Robotic system setups that are used in the experiments are introduced in Chapter V. All the master and slave systems are designed to have certain levels of fault tolerance. Hence, teleoperation systems configured in this study have fault tolerance features which enhance system reliability. This encourages the designer to use such systems in critical tasks.

One of the subsystems of teleoperation, the communications line, is not always dependable. There can always be time delays in communications line due to the distance involved between the remote system and the operator. Data loss during operation is another source of failure. These affect stability and tracking performance of the system. The teleoperation controllers to overcome these effects are presented in Chapter VI. Among these controllers, the wave variable technique is widely used in stabilizing teleoperation systems

197

with constant time delays. Two components added to this controller have increased the efficiency of this controller under variable time delays and communication or data losses. Finally, position/force controllers and their modifications are proposed as alternative master or slave controllers in teleoperation subsystems. The modifications on the architecture of these controllers have made them more dependable under time delays and communication losses.

Numerical simulations are carried out for single- and multiple-DOF teleoperation systems. The first set of simulations verified the necessity of wave variable technique under constant time delays for both single- and multiple-DOF systems. Various magnitudes of time delays are tested in these simulations. It is seen that as the time delays increase, the manipulation speed is required to be reduced not to damage the slave or its environment due to late response times. Variable time delays are then tested for a multiple-DOF system. It is observed that the customary wave variable technique is not sufficient enough to stabilize the system under variable time delays. The adaptive gain component introduced in Chapter VI is examined to stabilize the system. This modification produced satisfactory results and the system became stable. Later, position/force controllers are tested in numerical simulation studies using the Epson SCARA manipulator as the slave system. Among the controllers presented in Chapter VI, modified admittance controller is judged to be the most dependable for use in systems experiencing communication problems. The tests are carried out using this controller and produced satisfactory results in terms of applying the desired forces while achieving the motion demands without tracking problems, instabilities or chattering.

Experimental results are investigated for both limited- and unlimited-workspace teleoperation systems in Chapters VIII and IX. Various combinations of master and slave systems are used in each experiment. The limited-workspace teleoperation studies are initiated with the experiments using an identical master-slave teleoperation system. This teleoperation is composed of the gimbal-based master joystick and its virtual replica. Later, redundant slaves are used to configure redundant teleoperation systems. The first redundant teleoperation system is composed of the gimbal-based joystick and the virtual Fanuc arm. The second system is configured as the Phantom Omni Device as the master and the virtual model of the Motoman UPJ arm as the slave system. Commercial joystick from Logitech and the Phantom Omni Device is used as the master and slave systems for the last experimental setup.

The experiments are conducted under constant and variable time delays to observe the tracking performances and the stability. In experiments with constant time delays, the system experienced communication losses and inactivation of the wave variable technique for limited periods. The inactivation of wave variables simulated a case where the teleoperation starts with an unstable system. Although the customary wave variable technique was able to stabilize the system after these failures, position tracking errors are observed. As the position feedforward component proposed in Chapter VI is used, the new controller compensated for position offsets and satisfactory results are obtained for limited-workspace teleoperation systems under time delays. However, when the systems are exposed to variable time delays, this controller could not guarantee system stability. Hence, the adaptive gain component also presented in Chapter VI is added to the control scheme. Force tracking performance, on the other hand, was not affected in any of the experiments and therefore, a force feedforward component was found to be unnecessary in the controller architecture. The results were satisfactory as both components are used for teleoperation systems examined under variable time delays.

Later in Chapter VIII, the experimental results for position/force controllers are given. The experimental setup is composed of the gimbal-based joystick as the master system and the virtual Epson SCARA manipulator. The customary versions of the admittance and hybrid position/force controllers use the assumption that the error between the position demand and the actual position in Cartesian space is small. This assumption does not hold for teleoperation systems experiencing time delays and/or communication losses. Therefore, the architectures of these controllers are modified not to use this assumption by making the necessary comparisons in joint space rather than Cartesian space. This formulation holds for most of the industrial robots with valid inverse kinematics solutions.

Although both controllers are modified to be able to work in teleoperation systems experiencing time delays and/or communication losses, the modified hybrid position/force controller still requires a pure position controller to approach the contact surface. This type of control produces chattering especially when the system works under time delays or the communication is lost for a limited period. In contrast, admittance controller does not require switching between the controllers, but it stays active independent of the contact condition. Therefore, among the controllers investigated, it is concluded that this controller is the most suitable and dependable one for teleoperation systems. The experiments also proved that this controller allows the motion and force trajectories followed stably and without significant errors or any chattering. As a result of this, the modified admittance controller is suggested for the master and/or slave systems to produce reliable results when the system experiences time delays or communication losses.

Similar studies for stabilization and enhancing tracking performance are conducted for unlimited-workspace teleoperation systems under constant and variable time delays. Two different teleoperation systems are used in these experiments. The first teleoperation system consists of the gimbal-based master joystick and the virtual model of the holonomic mobile platform. The second system is configured as the commercial joystick as the master and the actual holonomic mobile platform as the slave system. The tracking priority is usually given to the velocity demand when unlimited-workspace slaves are used. Therefore, the experiments with the customary wave variables provided satisfactory results even after a limited period of communication loss or when it is applied to an unstable system under constant time delays. There were no significant errors in both velocity and force tracking performances in these experiments. The teleoperation systems are then tested under variable time delays. The customary wave variable technique was not successful to stabilize the system under the new conditions. However, the addition of an adaptive gain component is shown to stabilize the system even under variable time delays. As a result of this, the wave variable technique with adaptive gain component is recommended for unlimited-workspace teleoperation systems that experience variable time delays.

In the experiments, a total of seven teleoperation systems are used that range from identical master-slave to redundant and mobile platform teleoperation. This demonstrates that the proposed controllers are general purpose and can be employed on a variety of teleoperation systems successfully.

Overall, the teleoperation system designer first has to classify the system either as a limited- or unlimited-workspace bilateral teleoperation. Following this, certain levels of fault tolerance may be introduced to the master and slave system design. This is crucial for critical teleoperation systems where the system needs to continue the task even when one or more of its components fail. Then, the communications line conditions should be evaluated as accurately as possible to see if the system will experience constant or variable time delays or there may be communication or data losses during the operation. As a result of this, more sophisticated versions of the wave variables technique can be used to configure a stable system without significant tracking errors. Also, position/force controllers can be employed for master and/or slave systems not to exert excessive amounts of uncontrolled forces during these communication failures. The position/force controller suggested in this study is the modified admittance controller that came out to be more reliable than the other controllers considered. Applying all these components to the teleoperation system design, the end result becomes a more reliable system that can be applied to a wider range of critical missions.

10.2. Future Work

The magnitude of time delays used in this study is consistent with the measured Internet delays. The range of delays should be increased in future work concentrating on large-scale time delays observed in teleoperation. This will provide better knowledge for space teleoperation or Mars/Lunar teleoperation missions that experience time delays in the range of hours. Although there are widely used predictors available to predict communication time delays, a study on development of new predictors can also be addressed that use new methods such as neural networks for more accurate estimation of the delays.

More in-depth experimental work to study time delay effects via Internet must be carried out for the proposed controllers. The new experimental setup should be configured to connect master and slave systems located at different cities and even continents. Thus, the proposed controllers will find other platforms to be examined.

More work must be carried out for more rigorous stability proofs of the proposed controllers, which form another area for future studies. Also, controller design work should be continued to improve the proposed position feedforward component for teleoperation of manipulators that do not have valid analytical inverse kinematics solutions.

Moreover, system architecture and interface design presented in this work can also be further addressed. For instance, higher sampling rates may be achieved by optimizing the interface codes for the master and slave systems. The system architecture of the gimbal-based joystick can be altered to add another DOF to the system so that it will become a three-DOF force-reflecting joystick. This will enable it to be used in a wider variety of teleoperation systems that require more DOF.

Servomotors with more power should be used for the holonomic mobile platform to increase the speed of telemanipulation. The surface and servomotor condition affects the navigation of the mobile platform and results in changes in its orientation. Therefore, a gyroscope must be included in the system to monitor the orientation variations.

The master and slave systems used in experiments had either sensor, joint, mechanism or combination of sensor-mechanism level of fault tolerance. Different combinations of fault-tolerance features should be examined for use in critical tasks. This will provide the knowledge of the type of fault tolerance required for certain missions.

Although seven different teleoperation systems are investigated in this book, this study can be extended for teleoperation of other robotic devices such as unmanned airplanes, submarines, surgical robots, etc. New application areas will bring new uncertainties to the teleoperation systems and studies can be directed to compensate for these uncertainties. Also, the actual systems should be used in the future experiments instead of the virtual slaves as it was the case for some of the experiments in this work. The results of such extended work will also lend itself to further evaluate the accuracy of the recently developed system modeling techniques presented in this book.

REFERENCES

[1]   Capek, K. (1921). "RUR Rossum's Universal Robots", Prague.

[2]   Asimov, J., and Asimov, I. (2002). "It's been a Good Life," Prometheus Books, Amherst, NY.

[3]   Goshozono, W.K.Y., Kawabe, T., Kinami, H., Tsumaki, M., Uchiyama, Y., Oda, M., and Doi, M., (2004). "Model-Based Space Robot Teleoperation of ETS-VII Manipulator," IEEE Transactions on Robotics and Automation, vol. 20, is. 3, pp 602-612.

[4]   Imaida, T., Yokokohji, Y., Doi, T., Oda, M., and Yoshikawa, T., (2004). "Ground-space bilateral teleoperation of ETS-VII robot arm by direct bilateral coupling under 7-s time delay condition," IEEE Transactions on Robotics and Automation, vol. 20, is. 3, pp. 499-511.

[5]   Huang, P., Liu, Z., Zhao, G., Xu, W., and Liang, B., (2007). "A Ground Teleoperation Experimental System of Space Robot using Hybrid Approach," Proceedings of the IEEE International Conference on Integration Technology (ICIT '07), pp. 593-598.

[6]   Cavusoglu, M.C., (2000). "Telesurgery and Surgical Simulation: Design Modeling, and Evaluation of Haptic Interfaces to Real and Virtual Surgical Environments," Ph.D. Dissertation, University of California at Berkeley, Berkeley, CA.

[7]   Butner, S.E., and Ghodoussi, M., (2003). "Transforming a Surgical Robot for Human Telesurgery," IEEE Transactions on Robotics and Automation, vol. 19, is. 5, pp. 818-824.

[8]   Matsumoto, Y., Katsura, S., and Ohnishi, K., (2007). "Dexterous Manipulation in Constrained Bilateral Teleoperation Using Controlled Supporting Point," IEEE Transactions on Industrial Electronics, vol. 54, is. 2, pp. 1113-1121.

[9]   Romano, J.M., Webster III, R.J., and Okamura, A.M., (2007). "Teleoperation of Steerable Needles," Proceedings of the IEEE International Conference on Robotics and Automation, pp. 934-939.

[10] Sitti, M., (2003). "Teleoperated and Automatic Control of Nanomanipulation Systems using Atomic Force Microscope Probes," Proceedings of the IEEE Conference on Decision and Control, vol. 3, pp. 2118-2123.

[11] Bennet, D., and Needles, A., (1997). "A New Control Concept for Commercially Available Telerobotic Manipulators," Proceedings of the American Nuclear Society 7th Topical Meeting on Robotics and Remote Systems.

[12] Jet Propulsion Laboratory, California Institute of Technology, California. Retrieved September 26, 2007 from http://www-robotics.jpl.nasa.gov.

[13] Foster-Miller, Inc., Massachusetts. Retrieved September 26, 2007 from http://www.foster-miller.com.

[14] Vertut, J., and Coiffet, P., (1985). "Robot Technology Volume 3A: Teleoperation and robotics evolution and development," Prentice Hall, Englewood Cliffs, NJ.

[15] Massimino, M., and Sheridan, T.B., (1994). "Teleoperator Performance with Varying Force and Visual Feedback," Human Factors, vol. 36, is. 1, pp. 145-157.

[16] Howe, R.D., and Kontarinis, D. (1992). "Task performance with a dexterous teleoperated hand system," Proceedings of the Telemanipulator Technology Conference, pp. 199-207.

[17] Hannaford, B., Wood, L., McAffee, D.A., and Zak, H., (1991). "Performance evaluation of a six-axis generalized force-reflecting teleoperator," IEEE Transactions on Systems, Man, and Cybernetics, vol. 21, is. 3, pp. 620-633.

[18] Tanner, A.N., and Niemeyer, G., (2006). "High-Frequency Acceleration Feedback in Wave Variable Telerobotics," IEEE-ASME Transactions on Mechatronics, vol. 11, is. 2, pp. 119-127.

[19] Batsomboon, P., and Tosunoglu, S., (2000). "A Survey of Telesensation and Teleoperation Technology with Virtual Reality and Force Reflection Capabilities," International Journal of Modeling and Simulation, vol. 20, is. 1, pp. 79–88.

[20] Anderson, R.J., and Spong, W., (1989). "Bilateral Control of Teleoperation with Time Delay," IEEE Transaction on Automation and Control, vol. 34, is.5, pp.494-501.

[21] Niemeyer, G., and Slotine, J., (1997). "Using Wave Variables for System Analysis & Robot Control," Proceedings of the IEEE International Conference on Robotics and Automation, vol. 2, 1619-1625.

[22] Niemeyer, G., (1996). "Using Wave Variables in Time Delayed Force Reflecting Teleoperation," Ph.D. Dissertation, Massachusetts Institute of Technology, Cambridge, MA.

[23] Munir, S., and Book, W., (2002). "Internet-Based Teleoperation Using Wave Variables with Prediction," IEEE/ASME Transactions on Mechatronics, vol. 7, is. 2, pp. 124-133.

[24] Munir, S., (2001). "Internet-Based Teleoperation," Ph.D. Dissertation, Georgia Institute of Technology, Atlanta, GA.

[25] Munir, S., and Book, W., (2003). "Control Techniques and Programming Issues for Time Delayed Internet Based Teleoperation," Journal of Dynamic Systems, Measurement, and Control, vol. 125, is. 2, pp. 157-277.

[26] Chopra, N., Spong, M.W., Hirche, S., and Buss, M., (2003). "Bilateral Teleoperation over the Internet: the Time Varying Delay Problem," Proceedings of the American Control Conference, vol. 1, pp. 155-160.

[27] Hirche, S., (2002). "Control of Teleoperation Systems in QoS Communication Networks," Ph.D. Dissertation, Technical University of Berlin, Berlin, Germany.

[28] Chopra, N., Spong, M.W., Ortega, R., and Barabanov, N.E., (2004). "On Position Tracking in Bilateral Teleoperation," Proceedings of the American Control Conference, vol. 6, pp. 5244-5249.

[29] Griffin, W.B., (2003). "Shared Control for Dexterous Telemanipulation with Haptic Feedback," Ph.D. Dissertation, Stanford University, Palo Alto, CA.

[30] Lee D., Spong, M.W., and Martinez-Palafox, O., (2005). "Bilateral Teleoperation of Multiple Cooperative Robots with Delayed Communication: Theory," Proceedings of the IEEE International Conference on Robotics and Automation, pp. 368-373.

[31] Lawrence, D.A., (1993). "Stability and transparency in bilateral teleoperation," IEEE Transactions on Robotics and Automation, vol. 9, is. 5, pp. 624 -637.

[32] Hashtrudi-Zaad, K., and Salcudean, S.E., (2002). "Transparency in Time-Delayed Systems and the Effect of Local Force Feedback for Transparent Teleoperation," IEEE Transactions on Robotics and Automation, vol. 18, is. 1, pp. 108-114.

[33] Delwiche, T., and Kinnaert, M., (2007). "A Four-Channel Adaptive Structure For High Friction Teleoperation Systems In Contact With Soft Environments," Proceedings of the IEEE International Conference on Robotics and Automation, pp. 1631-1636.

[34] Hashtrudi-Zaad, K., and Salcudean, S.E., (2001). "Analysis of Control Architectures for Teleoperation Systems with Impedance/Admittance Master and Slave Manipulators," The International Journal of Robotics Research, vol. 20, is. 6, pp. 419-445.

[35] Aziminejad, A., Tavakoli, M., Patel, R.V., and Moallem M., (2007). "Wave-Based Time Delay Compensation in Bilateral Teleoperation: Two-Channel versus Four-Channel Architectures," Proceedings of the American Control Conference, pp. 1449-1454.

[36] Kitts, C., Quinn, N., Ota, J., Stang, P., and Palmintier, B., (2003). "Development and Teleoperation of Robotic Vehicles," Proceedings of the 2nd AIAA "Unmanned Unlimited" Conference and Workshop and Exhibit.

[37] Cavallaro, J.R., Near, C.D., and Uyar, M.U., (1989). "Fault-Tolerant VLSI Processor Array for the SVD," Proceedings of the IEEE International Conference on Computer Design, pp. 176-180.

[38] Kieckhafer, R.M., Walter, C.J., Finn, A.M., and Thambidurai, P.M., (1988). "MAFT Architecture for Distributed Fault Tolerance," IEEE Transactions on Computers, vol. 37, is. 4, pp. 398-405.

[39] Merrill, W.C., DeLaat, J.C., and Bruton, W.M., (1988). "Advanced Detection, Isolation, and Accommodation of Sensor Failures - Real-Time Evaluation," Journal of Guidance, Control, and Dynamics, vol.11, is. 6, pp. 517-526.

[40] Tesar, D., Sreevijayan, D., and Price, C., (1990). "Four-Level Fault Tolerance in Manipulator Design for Space Operations," Proceedings of the First International Symposium on Measurement and Control in Robotics.

[41] Visinsky, M.L., (1991). "Fault Detection and Fault Tolerance Methods for Robotics," Master's Thesis, Rice University, Houston, TX.

[42] Nelson, V.P., (1990). "Fault-Tolerant Computing: Fundamental Concepts," IEEE Computer, vol. 23, is. 7, pp. 19-25.

[43] Sklaroff, J.R., (1976). "Redundancy Management Technique for Space Shuttle Computers," IBM Journal of Research and Development, vol. 20, is. 1, pp. 20-28.

[44] Norman, P.G., (1987). "The New AP101S General-Purpose Computer (GPC) for the Space Shuttle," Proceedings of the IEEE Special Issue on Progress in Space - From Shuttle to Station, vol. 75, is. 3, pp. 308-319.

[45] Wu, E., Diftler, M., Hwang, J., and Chladek, J., (1991). "A Fault Tolerant Joint Drive Systems for the Space Shuttle Remote Manipulator System," Proceedings of the IEEE International Conference on Robotics and Automation, pp. 2504-2509.

[46] Hamilton, D.L., Visinsky, M.L., Bennett, J.K., Cavallaro, J.R., and Walker, I.D., (1994). "Fault Tolerant Algorithms and Architectures for Robotics," Proceedings of the 7th Mediterranean Electrotechnical Conference, vol. 3, pp. 1034-1036.

[47] Maciejewski, A.A., (1990). "Fault Tolerant Properties of Kinematically Redundant Manipulators," Proceedings of the IEEE Conference on Robotics and Automation, vol. 1, pp. 638-642.

[48] Deo, A.S., (1992). "Robot Subtask Performance with Singularity Robustness using Optimaldamped Least Squares," Proceedings of the IEEE International Conference on Robotics and Automation, vol. 1, pp. 434-441.

[49] Tosunoglu, S., (1995). "Intelligent Control Systems for Fault-Tolerant Manipulators," Recent Advances in Mechatronics, Editors O. Kaynak, M. Ozkan, N. Bekiroglu, and I. Tunay, Vol. 1, pp. 356–362, Bogazici University Publication, Istanbul, Turkey.

[50] Hamilton, D.L., Bennett, J.K., and Walker, I.D., (1992) "Simulation of a reliable robot control architecture," Proceedings of International Simulation Technology Conference, pp. 321-327.

[51] Spooner, J.T., and Passino, K.M., (1997). "Fault-Tolerant Control for Automated Highway Systems," IEEE Transactions on Vehicular Technology, vol. 46, is. 3, pp. 770-785.

[52] Stanley-Marbell, P., and Marculescu, D., (2004). "Local Decisions and Triggering Mechanisms for Adaptive Fault-Tolerance," Proceedings of IEEE Design, Automation and Test in Europe Conference, vol. 2, pp. 968-973.

[53] Diao, Y., and Passino, K.M., (2001). "Stable Fault-Tolerant Adaptive Fuzzy/Neural Control for a Turbine Engine," IEEE Transactions on Control Systems Technology, Vol. 9, No. 3, pp. 494-509.

[54] Monteverde, V., and Tosunoglu, S., (2000). "Comparison of Robot Control Strategies for Fault-Tolerant Manipulators," Proceeding of the IEEE 32nd Southeastern Symposium on System Theory, pp. 99–103.

[55] Dede, M., and Tosunoglu, S., (2006). "Fault-Tolerant Teleoperation Systems Design," Industrial Robot: An International Journal, vol. 33, is. 5, pp. 365-372.

[56] Matlab©, The MathWorks, Inc. Retrieved September 26, 2007 from http://www.mathworks.com.

[57] Alliance Spacesystems, LLC. Retrieved September 26, 2007 from http://www.alliancespacesystems.com.

[58] SolidWorks Corporation. Retrieved September 26, 2007 from http://www.solidworks.com.

[59] Commuter Cars Cororation. Retrieved September 26, 2007 from http://www.commutercars.com.

[60] Pap, J-S., Xu, W., and Bronlund, J., (2005). "A robotic human masticatory system: kinematics simulations," International Journal of Intelligent Systems Technologies and Applications, vol.1, is. 1/2, pp. 3-17.

[61] Omura, Y., (2006). "Robotic software simulator: aids robotic development (Real motion system software)," Design News, vol. 61, is. 4, pp. 56-58.

[62] Madadi, V., Dede, M.I.C., and Tosunoglu, S., (2007). "Gait Development for the Tyrol Biped Robot," ASME Early Career Technical Journal, vol. 6, is. 1, pp. 7.1-7.8.

[63] Dede, M.I.C., Nasser, S., Ye, S., and Tosunoglu, S., (2006). "Cerberus: Development of a Humanoid Robot," ASME Early Career Technical Journal, vol. 5, is. 1, pp. 8.1-8.8.

[64] Boeing. Retrieved September 26, 2007 from http://www.boeing.com.

[65] Daimler AG. Retrieved September 26, 2007 from http://www.daimler.com.

[66] Lockheed Martin Space Systems Company. Retrieved September 26, 2007 from http://www.lockheedmartin.com/ssc.

[67] Dede, M.I.C., and Tosunoglu, S., (2007). "Development of a Virtual Force-Reflecting SCARA Robot for Teleoperation," Innovative Algorithms and Techniques in Automation, Industrial Electronics and Telecommunications, T. Sobh, K. Elleithy, A. Mahmood, M. Karim (Eds.), pp. 293-298, Springer, Dordrecht, The Netherlands.

[68] Connolly, T.J., Blackwell, B.B., and Lester, L.F., (1989). "A simulator-based approach to training in aeronautical decision making," Aviation Space Environment Medicine, vol. 60, is. 1, pp. 50-52.

[69] Page, E.H., and Smith, R., (1998). "Introduction to Military Training Simulation: A Guide for Discrete Event Simulationists," Proceedings of the Winter Simulation Conference, pp. 53-60.

[70] Dinsmore, M., Langrana, N., Burdea, G., and Ladeji, J., (1997). "Virtual reality training simulation for palpation of subsurfacetumors," Proceedings of the IEEE Virtual Reality Annual International Symposium, pp. 54-60.

[71] Seymour, N.E., Gallagher, A.G., Roman, S.A., O'Brien, M.K., Bansal, V.K., Andersen, D.K., and Satava, R.M., (2002). "Virtual Reality Training Improves Operating Room Performance: Results of a Randomized, Double-Blinded Study," Annals of Surgery, vol. 236, is. 4, pp. 458-464.

[72] Hays, R.T., Jacobs, J.W., Prince, C., and Salas, E., (1992). "Flight Simulator Training Effectiveness: A Meta-Analysis," Military Psychology, vol. 4, is. 2, pp. 63-74.

[73] Bell, H.H., and Waag, W.L., (1998). "Evaluating the Effectiveness of Flight Simulators for Training Combat Skills: A Review," International Journal of Aviation Psychology, vol. 8, is. 3, pp. 223-242.

[74] Guo, K., Guan, H., and Zong, C., (1999). "Development and applications of JUT-ADSL driving simulator," Proceedings of the IEEE International Vehicle Electronics Conference, pp. 1-5.

[75] Jorgensen, C., Wheeler, K., and Stepniewski, S., (2000). "Bioelectric Control of a 757 Class High Fidelity Aircraft Simulation," Proceedings of the 3rd International Symposium on Intelligent Automation and Control, pp.1-8.

[76] Miner, N.E., and Stansfield, S.A., (1994). "An Interactive Virtual Reality Simulation System for Robot Control and Operator Training," Proceedings of the IEEE International Conference on Robotics and Automation, Volume 2, pp. 1428-1435.

[77] Ziegler, R., Fischer, G., Müller, W., and Göbel, M., (1995). "Virtual Reality Arthroscopy Training Simulator," Computers in Biology and Medicine, vol. 25, is. 2, pp. 193-203.

[78] Kang, H.S., Jalil, M.K.A., and Mailah, M., (2004). "A PC-based Driving Simulator Using Virtual Reality Technology," Proceedings of the ACM SIGGRAPH International Conference on Virtual Reality Continuum and Its Applications in Industry, pp. 273-277.

[79] Lee, W.S., Kim, J.H., and Cho, J.H., (1998). "A Driving Simulator as a Virtual Reality Tool," Proceedings of the IEEE International Conference on Robotics and Automation, pp. 71-76.

[80] Dede, M.I.C., and Tosunoglu, S., (2006). "Development of a Real-Time Force-Reflecting Teleoperation System Based on Matlab© Simulations," Proceedings of the 19th Florida Conference on Recent Advances in Robotics.

[81] Genius, KYE Systems Corp. Retrieved September 26, 2007 from http://www.geniusnet.com.

[82] Logitech. Retrieved September 26, 2007 from http://www.logitech.com.

[83] SensAble Technologies, Inc. Retrieved September 26, 2007 from http://www.sensable.com.

[84] Dede, M.I.C., and Tosunoglu, S., (2006). "Design of a Fault-Tolerant Holonomic Mobile Platform," Proceedings of the 19th Florida Conference on Recent Advances in Robotics.

[85]   Dr. Robot, Inc. Retrieved September 26, 2007 from http://www.drrobot.com.

[86]   Motoman Inc. Retrieved September 26, 2007 from http://www.motoman.com.

[87]   AMETEK, Inc. Retrieved September 26, 2007 from http://www.pittmannet.com.

[88]   Galil Motion Control Inc. Retrieved September 26, 2007 from http://www.galilmc.com.

[89]   Li, Y., Harari, S., Wong, H., and Kapila, V., (2004). "Matlab-Based Graphical User Interface Development for Basic Stamp 2 Microcontroller Projects," Proceedings of the American Control Conference, vol.4, pp. 3233-3238.

[90]   EPSON Robots. Retrieved September 26, 2007 from http://www.robots.epson.com.

[91]   FANUC LTD. Retrieved March 30, 2007 from http://www.fanucrobotics.com.

[92]   Marshall, J.E., (1979). "Control of Time-Delay Systems," The Institution of Electrical Engineers, London and New York, Peter Peregrinus Ltd., Stevenage, UK.

[93]   Kuo, B.C., (1991). "Automatic Control Systems," Prentice Hall, Englewood Cliffs, NJ.

[94]   Truxal, J.G., (1955). "Automatic Feedback Control System Synthesis," McGraw-Hill Book Company, New York, NY.

[95]   Cho, H.C., and Park, J.H., (2002). "Impedance Controller Design of Internet-Based Teleoperation Using Absolute Stability Concept," Proceedings of IEEE/RSJ International Conference on Intelligent Robots and Systems EPFL, vol.3, pp. 2256-2261.

[96]   Watanabe, K., and Ito, M., (1981). "An observer for linear feedback control laws of multivariable systems with multiple delays in controls and outputs," Systems and Control Letters, vol.1, is.1, pp. 54-59.

[97]   Brady, K., and Tarn, T.J., (1998). "Internet-Based Remote Teleoperation," Proceedings of the IEEE International Conference on Robotics and Automation, vol. 1, pp. 65-70.

[98]   Park, J.H., and Cho, H.C., (2000). "Sliding Mode Control of Bilateral Teleoperation Systems with Force-Reflection on the Internet," Proceedings of IEEE/RSJ International Conference on Intelligent Robots and Systems, pp. 1187-1192.

[99]   Alise, M., Roberts, R.G., and Repperger, D.W., (2006). "The Wave Variable Method for Multiple Degree of Freedom Teleoperation Systems with Time Delay," Proceedings IEEE International Conference on Robotics and Automation, pp. 2908-2913.

[100] Chopra, N., Spong, M.W., Ortega R., and Barbanov, N.E., (2006). "On Tracking Performance in Bilateral Teleoperation", IEEE Transaction on Robotics, vol. 22, is. 4, pp. 861-866.

[101] Dede M.I.C., and Tosunoglu, S., (2007). "Modification of the Wave Variable Technique for Teleoperation Systems Experiencing Communication Loss," Proceedings of the 7th IEEE International Symposium on Computational Intelligence in Robotics and Automation, pp. 380-385.

[102] M.K. Özgoren, (2002). "Topological Analysis of 6-Joint Serial Manipulators and Their Inverse Kinematics Solutions," Mechanism and Machine Theory, vol. 37, is. 5, pp. 511-547.

[103] Chopra, N., Spong, M.W., Ortega, R., and Barabanov, N.E., (2006). "On Tracking Performance In Bilateral Teleoperation," IEEE Transactions on Robotics, vol. 22, is. 4, pp. 861-866.

[104] Innovative Technology, (2001). "BOA II: Asbestos Pipe-Insulation Removal Robot System," Summary Report Prepared for U.S. Department of Energy Office of Environmental Management Office of Science and Technology, Washington, DC.

[105] Dede, M.I.C., (2003). "Position/Force Control of Robot Manipulators," M.Sc. Thesis, Middle East Technical University, Ankara, Turkey.

[106] Seraji, H., (1994). "Adaptive admittance control: an approach to explicit force control in compliant motion," Proceedings of the IEEE International Conference on Robotics and Automation, vol. 4, pp. 2705-2712.

[107] Raibert, M.H., and Craig J.J., (1981). "Hybrid position/force control of manipulators," ASME Journal of Dynamic Systems, Measurements, and Control, vol. 103, is. 2, pp. 126-133.

[108] Zeng, G., and Hemami, A., (1997). "An overview of robot force control," Robotica, vol. 15, pp. 473-482.

[109] Dede, M.I.C., and Ozgoren, M.K., (2004). "A new approach for the formulation of the admittance and hybrid position/force control schemes for industrial manipulators," Proceedings of the 10th Robotics and Remote Systems Meeting, pp. 332-337.

[110] Duffy, J., (1980). "Analysis of Mechanisms And Robot Manipulators" Wiley, New York, NY.

[111] Dede, M.I.C., and Tosunoglu, S., (2004). "Modeling of a Single DOF Force Feedback Teleoperation System," Proceedings of the 17th Florida Conference on Recent Advances in Robotics.

[112] Dede, M.I.C., Tosunoglu, S., and Repperger, D., (2004). "Effects of Time Delay on Force-Feedback Teleoperation Systems," Proceedings of the 12th Mediterranean Conference on Control and Automation.

[113] Dede, M.I.C., Tosunoglu, S., and Repperger, D., (2004). "A Study on Multiple Degree-of-Freedom Force Reflecting Teleoperation" Proceedings of the International Conference on Mechatronics, pp. 476-481.

[114] Dede, M.I.C., Tosunoglu, S., and Repperger, D., (2005). "A Study on Multiple Degree-of-Freedom Force-Reflecting Teleoperation with Constant and Variable Time Delays," ASME Southeastern Region XI Technical Journal, vol. 4, is.1, pp. 3.1-3.8.

[115] Dede, M.I.C., and Tosunoglu, S., (2007). "Parallel Position/Force Controller for Teleoperation Systems," IFAC Workshop on Technology Transfer in Developing Countries: Automation in Infrastructure Creation – TT, DECOM-TT.

[116] Dede, M.I.C., and Tosunoglu, S., (2007). "Modification of the Wave Variable Technique for Teleoperation Systems Experiencing Communication Loss," Proceedings of the IEEE International Symposium on Computational Intelligence in Robotics and Automation, pp. 380-385.

[117] Dede, M.I.C., and Tosunoglu, S., (2007). "Control of Teleoperation Systems Experiencing Communication Loss," Robotics and Automation Systems, Submitted.

[118] Dede, M.I.C., and Tosunoglu, S., (2007). "Tracking Performance of an Identical Master-Slave Teleoperation System under Variable Time Delays," International Joint Conferences on Computer, Information and Systems Sciences, and Engineering (CISSE 2007 Online International E-Conference), and International Conference on Industrial Electronics, Technology and Automation, (IETA 2007), Submitted.

[119] Dede, M.I.C., and Tosunoglu, S., (2007). "Position Tracking Performance of a Redundant Teleoperation," Proceedings of the 20th Florida Conference on Recent Advances in Robotics.

[120] Dede, M.I.C., and Tosunoglu, S., (2007). "Parallel Position/Force Controller for Teleoperation Systems with Time Delays," International Journal of Robotics and Automation, Submitted.

[121] West, M., and Asada, H.H., (1997). "Design of Ball wheel Mechanisms for Omnidirectional Vehicles with Full Mobility and Invariant Kinematics," Journal of Mechanical Design, vol.119, is. 2, pp.153-161.

[122] Kornylak Corp. Retrieved March 15, 2006 from http://www.kornylak.com.

[123] Wilson, L., Williams, C., Yance, J., Lew, J., Williams II, R.L., and Gallina, P., (2001). "Design and Modeling of a Redundant Omni-directional RoboCup Goalie," Proceedings of the RoboCup 2001 International Symposium.

[124] Carter, B., Good, M., Dorohoff, M., Lew, J., Williams II, R. L., and Gallina, P., (2001). "Mechanical Design and Modeling of an Omni-directional RoboCup Player," Proceedings of the RoboCup 2001 International Symposium.

[125] NARP All Side Rollers. Retrieved March 15, 2006 from http://www.narp-trapo.com.

[126] Blackwell, M., (1990). "The URANUS Mobile Robot," Carnegie Mellon University Technical Report, CMU-RI-TR-91-06.

[127] Acroname Inc. Retrieved in March 15, 2006 from http://www.acroname.com.

[128] Purwin, O., and D'Andrea, R., (2006). "Trajectory Generation and Control for Four Wheeled Omnidirectional Vehicles," Robotics and Autonomous Systems, vol. 54, is. 1, pp.13-22.

[129] Wada, M., and Asada, H.H., (1999). "Design and Control of a Variable Footprint Mechanism for Holonomic Omnidirectional Vehicles and its Application to Wheelchairs," IEEE Transactions on Robotics and Automation, vol. 15, is. 6, pp. 978-989.

[130] Yu, H., Spenko, M., and Dubowsky, S., (2004). "Omni-Directional Mobility Using Active Split Offset Castors," Journal of Mechanical Design, vol.126, pp. 822-829.

[131] Holmberg, R., and Khatib, O., (2000). "Development and Control of a Holonomic Mobile Robot for Mobile Manipulation Tasks," International Journal of Robotics Research, vol. 19, is. 11, pp. 1066-1074.

[132] Xiao, J., (2006). "Mobot: Mobile Robot," In Introduction to Robotics Lecture Notes, Department of Electrical Engineering, City College of New York, NY, Retrieved in March 2006 from www-ee.ccny.cuny.edu/www/web/jxiao/mobot.pdf, accessed March 15, 2006.

[133] Parallax, Inc. Retrieved in March 15, 2006 from http://www.parallax.com.

[134] Hitec RCD Usa, Inc. Retrieved in March 15, 2006 from http://www.hitecrcd.com.

[135] Futaba, Inc. Retrieved in March 15, 2006 from http://www.futaba-rc.com.

[136] Pontech, Inc. Retrieved in March 15, 2006 from http://www.pontech.com.

[137] New Micros, Inc. Retrieved in March 15, 2006 from http://www.newmicros.com.

[138] Gridconnect, Inc. Retrieved in March 15, 2006 from http://www.ipenabled.com.

APPENDICES

APPENDIX A – INTERFACE SOURCE CODE FOR GIMBAL-BASED JOYSTICK

APPENDIX B – DESIGN STAGES OF THE HOLONOMIC MOBILE PLATFORM

APPENDIX C – INTERFACE SOURCE CODES FOR THE HOLONOMIC MOBILE
PLATFORM

APPENDIX A

INTERFACE SOURCE CODE FOR GIMBAL-BASED JOYSTICK


```
//This is a C++ code created to be an interface between the Matlab Simulink environment and Galil Motion
control Card
//First necessary libraries are included to the code including Galil Motion Libraries
//Necessary commands are used to make this code used as an S-function block in Matlab
//Number of input and output ports of this S-function block is assigned
//Force demands are received from the slave through Matlab and the input ports of this S-function
//Encoder readings are collected from the motion control card
//Force commands received from Matlab is compared with servomotor position to drive the servos
//Torque commands to drive the servomotors is collected in two strings for both axes
//Torque commands are sent to the motion control card to drive the servomotors
//Joystick positions about each axes is sent to Matlab through the output ports of the S-function

//Library is included in this portion of the code
#include "stdafx.h"
#include <windows.h>
#include <math.h>
#include "dmccom.h" //This header file is from Galil Motion Control
#include "stdio.h"
#include <iostream>
#include <string>
using std::string;
#include <sstream>
#include <stdlib.h>
using namespace std;

long rc;
HANDLEDMC hDmc; //These are functions to be used to operate with Galil Motion Controller Card
HWND hWnd;

#define S_FUNCTION_NAME  sdof_v4 //This portion is to create S-function block in Matlab
#define S_FUNCTION_LEVEL 2
#define HUGE_VAL

#include <simstruc.h>
int xaxis;
int yaxis;

char str[100];
float FrcX;
float FrcY;

//Number of Input and Output Ports are declared for the S-function block in the portion below
static void mdlInitializeSizes(SimStruct *S)
{
    ssSetNumSFcnParams(S, 0);
    if (ssGetNumSFcnParams(S) != ssGetSFcnParamsCount(S)) {
      return;
    }

    if (!ssSetNumInputPorts(S, 2)) return;
    ssSetInputPortWidth(S, 0, DYNAMICALLY_SIZED);
        ssSetInputPortWidth(S, 1, DYNAMICALLY_SIZED);

    ssSetInputPortDirectFeedThrough(S, 0, 1);
```

```
                ssSetInputPortDirectFeedThrough(S, 1, 1);

        if (!ssSetNumOutputPorts(S,2)) return;
        ssSetOutputPortWidth(S, 0, DYNAMICALLY_SIZED);
                ssSetOutputPortWidth(S, 1, DYNAMICALLY_SIZED);

        ssSetNumSampleTimes(S, 1);

        ssSetOptions(S,
                SS_OPTION_WORKS_WITH_CODE_REUSE |
                SS_OPTION_EXCEPTION_FREE_CODE |
                SS_OPTION_USE_TLC_WITH_ACCELERATOR);
}

static void mdlInitializeSampleTimes(SimStruct *S)
{
        ssSetSampleTime(S, 0, INHERITED_SAMPLE_TIME);
        ssSetOffsetTime(S, 0, 0.0);
}

static void mdlOutputs(SimStruct *S, int_T tid)
{

InputRealPtrsType uPtrs = ssGetInputPortRealSignalPtrs(S,0);
real_T          *y1  = ssGetOutputPortRealSignal(S,0);
real_T          *y2  = ssGetOutputPortRealSignal(S,1);
int_T           width = ssGetOutputPortWidth(S,0);

//Force commands are received from the slave through Matlab and input ports of this s-function block
real_T FX = (float)(*uPtrs[0]);
real_T FY = (float)(*uPtrs[1]);

//The below portion of the code is required for data collection from the encoders of the servomotors
DMCDATARECORDQR MyDataRecordQR;
rc = DMCOpen(1, hWnd, &hDmc);

char szBuffer[64];

rc = DMCCommand(hDmc, "QR\r", (LPCHAR)&MyDataRecordQR,
sizeof(MyDataRecordQR));

xaxis=MyDataRecordQR.DataRecord.AxisInfo[2].MotorPosition; //Encoder data is recorded to xaxis
yaxis=MyDataRecordQR.DataRecord.AxisInfo[1].MotorPosition; //Encoder data is recorded to yaxis

FrcX=1;
FrcY=1;

//Force commands received from Matlab is compared with servomotor position to drive the servos
if ((FX > 0 && xaxis < 0) || (FX < 0 && xaxis > 0))
                {
                FrcX=1+fabs(FX);
                }
if ((FY > 0 && yaxis < 0) || (FY < 0 && yaxis > 0))
                {
                FrcY=1+fabs(FY);
                }
//Torque commands to drive the servomotors is collected in two strings for both axes
stringstream buff;
string forcex;
buff << FrcX;
```

```cpp
buff >> forcex;
string torkx = "TLZ=" + forcex + ';';

stringstream buff1;
string forcey;
buff1 << FrcY;
buff1 >> forcey;
string torky = "TLY=" + forcey + ';';

char *torqx;
char *torqy;

torqx = const_cast<char *>(torkx.c_str());
torqy = const_cast<char *>(torky.c_str());

//sends torque commands to drive x-axis servomotor
rc = DMCCommand(hDmc, torqx, szBuffer, sizeof(szBuffer));

//sends torque commands to drive y-axis servomotor
rc = DMCCommand(hDmc, torqy, szBuffer, sizeof(szBuffer));

*y1=xaxis; //sends joystick position about x-axis to Matlab
*y2=yaxis; //sends joystick position about x-axis to Matlab

}

static void mdlTerminate(SimStruct *S)
{
}
//Last commands to change to C code into an S-function block
#ifdef  MATLAB_MEX_FILE
#include "simulink.c"
#else
#include "cg_sfun.h"
#endif
```

APPENDIX B

DESIGN STAGES OF THE HOLONOMIC MOBILE PLATFORM

B.1 Mobile Platform Background

Mobile robots are either designed with legs, wheels or have a combination of these two. The wheels used to build mobile platforms vary in design. Some of these wheels are called fixed wheel, centered orientable, off-centered orientable (caster), and omni-directional (Swedish) wheel as shown in Figure B.1.

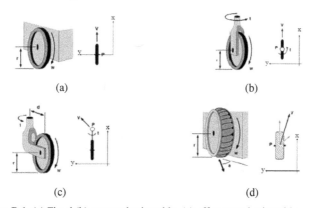

(a)   (b)

(c)   (d)

Figure B.1. (a) Fixed (b) centered orientable, (c) off-centered orientable, (d) omni-directional wheel

Omni-directional wheel is generally used in building holonomic mobile platforms. These platforms are capable of moving in any direction at any orientation. Their orientation can also be changed without affecting the linear motion along the other axes. Thus, motion along all three degree-of-freedoms of the mobile platform achieved independently in an uncoupled fashion.

There are different designs for omni-directional wheels. West and Asada [121] developed a ball wheel mechanism to be used as an omni-directional wheel. In the ball wheel design, power from the motor is transmitted through gears to an active roller ring and then to the ball via friction between the rollers and the ball as shown in Figure B.2.

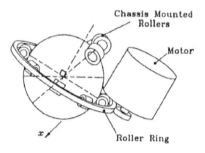

Figure B.2. Ball wheel mechanism [1]

Kornylak Corporation manufactures two types of omni-directional wheels as shown in Figure B.3 [122]. Researchers at Ohio University have used the Transwheel® of Kornylak Corp. in their design of omni-directional RoboCup players and goalkeeper [123, 124].

(a)   (b)

Figure B.3. (a) Transwheel®, (b) Omniwheel of Kornylak Corp. [122]

North American Roller Products (NARP) [125] also produces omni-directional wheels in different sizes and materials. The company calls these omni-directional wheels "All-side Rollers" as shown in Figure B.4.

Figure B.4. All-side Roller of North American Roller Products [125]

Scientists at Carnegie Mellon University built Uranus mobile robot in early 90's to provide general-purpose mobile base to support research in indoor robot navigation [126]. As illustrated in Figure B.5, Uranus had four traditional Swedish (Mecanum) wheels. Although the locations of the wheels are in customary formation for four wheeled vehicles, the unique design of the Swedish wheels enabled the platform to move independently in all three degrees-of-freedom.

Figure B.5. CMU's Uranus [126]

Three actuated omni-directional wheels are sufficient to develop a holonomic vehicle that moves independently in all three degrees-of-freedom. Researchers at Carnegie Mellon University's Robotics Institute have later commercialised a new design of the holonomic mobile platform named Palm Pilot Robotic Kit (PPRK) [127]. In this design, they used three omni-directional wheels from North American Roller Products in a triangular configuration as shown in Figure B.6.

Figure B.6. CMU's PPRK [127]

220

Omni-directional mobile platforms became very popular in RoboCup (robot soccer games) since 2000. Teams started building their players using omni-directional wheels in order to develop a holonomic mobile platform. The reason for this is that these platforms are more manoeuvrable, able to navigate tight quarters, and are easier to control. Participants from the Ohio University and Padova University developed two versions of omni-directional mobile platforms for RoboCup. The first version is developed to be a regular player in the robot soccer game and had same wheel configuration as PPRK as illustrated in Figure B.7 [123]. The second version was developed as a goalkeeper with four omni-directional wheels [124]. The design stage and the actual constructed version of this platform can be depicted in Figure B.8. Goalkeeper was designed to have redundancy so it will have a better mobility, specifically to be able to go sideways easily. As shown in Figure B.9, the Cornell University team also used a four-omni-directional wheeled platform for their RoboCup player design [128].

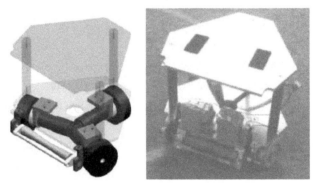

Figure B.7. RoboCup player of Ohio University team [123]

Figure B.8. RoboCup goalkeeper of Ohio University team [124]

221

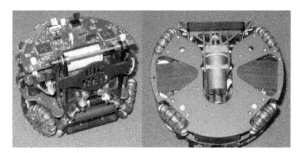

Figure B.9. RoboCup goalkeeper of Cornell University team [128]

Having developed the ball wheel mechanism, West and Asada used four of these mechanisms to develop an omni-directional base of a wheelchair in Figure B.10 [129]. This vehicle had four independent servomotors driving the four ball wheels that allow the vehicle to move in an arbitrary direction from an arbitrary configuration as well as to change the angle between the two beams and thereby change the footprint. They had three objectives for having the control of the beam angle. One is to augment static stability by varying the footprint so that the mass centre of the vehicle may be kept within the footprint at all times. The second is to reduce the width of the vehicle when going through a narrow doorway. The third is to change the gear ratio relating the vehicle speed to individual actuator speeds.

Some researchers chose to work with caster wheels instead of omni-directional wheels to develop holonomic mobile platforms. Yu, Spenko, and Dubowsky developed a mobile platform which they called SmartWalker (Figure B.11) using two active split offset casters (ASOC) and a conventional caster [130].

Figure B.10. Reconfigurable omni-directional mobile platform [129]

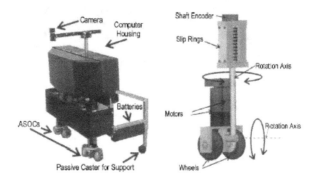

Figure B.11. SmartWalker and its active split offset caster [130]

Holmberg and Khatib were also developed their holonomic mobile platform using powered (active) castor. In their research, they used Nomadic XR4000 mobile platform (Figure B.12) with four powered caster wheels [131].

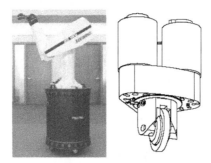

Figure B.12. Nomadic XR4000 and its powered caster model [131]

B.2. Design Criteria

The mobile platform to be designed will be used as a slave system in teleoperation experiments. Thus, it will not be an autonomous vehicle but will be driven by the commands received from the master system. It should also send sensory information back to the master system. The master system is the two-DOF gimbal-based joystick. The DOF of the joystick are uncoupled due to its unique gimbal design. An extra uncoupled degree-of-freedom can also be added to the joystick for future studies. Hence, the mobile platform should have three degrees-of-freedom to be compatible with the joystick and preferably they should be uncoupled.

223

Another requirement for this platform is to have fault tolerance in order to be used in fault-tolerant teleoperation systems. Since the robot to be designed is a mobile platform it should not be connected to the master system through cables. Finally, the desired features of the mobile platform are listed below:

- Three uncoupled degrees-of-freedom
- Sense the environment that it is working on
- Receive information from the master system
- Send information to the master system
- Cable-free communication with the master system
- Fault-tolerant design

B.3. Conceptual Designs

In the design process of the mobile platform, four design concepts are considered as briefly outlined below.

- Design Concept 1

The first design concept is a two-wheel rear-end drive with Ackerman steering system (Figure B.13). This design is mostly used in automobile industry. Fault tolerance is achieved by changing the design so that reserve servomotors are active to drive the wheels when there is a faulty servomotor. Its specifications are as follows:

Drive type: Two-wheel rear drive with Ackerman steering system

Total number of servos: 2

Degree-of-mobility: 1

Mobility: Travels in any direction by changing its orientation using its Ackerman steering system

Fault-tolerant design: Possible in joint level

This vehicle has fixed arc motion, which means it has only one Instantaneous Center of Rotation (ICR). Therefore, it has one degree-of-mobility.

Figure B.13. Two-wheel differential drive system [132]

224

- Design Concept 2

The second design concept is a two-wheeled differential drive system with a third point of contact by a roller-ball is used for balance (Figure B.14). Fault tolerance is achieved by changing the design so that reserve servomotors become active to drive the wheels when there is a faulty servomotor. Its specifications are as follows:

Drive type: Two-wheel differential

Total number of servos: 2

Degree of mobility: 2

Mobility: Travels in any direction by changing its orientation also can rotate about its wheels' midpoint

Fault-tolerant design: Possible in joint level

This vehicle has variable arc motion, which means it has a line of ICRs. Therefore, it has two degrees-of-mobility.

ICR

Figure B.14. Two-wheel differential drive system [132]

- Design Concept 3

The third design concept is a three-omni-directional wheel drive system (Figure B.15). This design is widely used in RoboCup player design in the past 6 years. It is a holonomic vehicle by design. Fault tolerance is achieved by changing the design so that reserve servomotors become active to drive the wheels when there is a faulty servomotor. Its specifications are summarized as follows:

Drive type: Three omni-directional wheel drive system

Total number of servos: 3

Degree-of-mobility: 3

Mobility: Travels in any direction at any orientation

Fault-tolerant design: Possible in joint level

225

This vehicle has fully free motion, which means ICRs can be located at any position. Therefore, it has three degrees-of-mobility.

Figure B.15. Three-omni-directional wheel drive system [132]

- Design Concept 4

The fourth design concept is a four-omni-directional wheel drive system (Figure B.16). This design is very similar to the third design concept, but it is redundant in link level. It is a holonomic fault-tolerant vehicle by design. Its specifications are listed below:

Drive type: Four omni-directional wheel drive system

Total number of servos: 4

Degree-of-mobility: 3

Mobility: Travels in any direction at any orientation

Fault-tolerant design: Link level fault tolerance by design

This vehicle has fully free motion, which means ICRs can be located at any position. Therefore, the platform has three degrees-of-mobility. Even if one of the wheels fails, the vehicle will still have three degrees-of-mobility.

Figure B.16. Four-omni-directional wheel redundant drive system

B.4. Final Design

All the design concepts can be built to exchange information with the master system via cable-free communication while gathering information about the environment using sensors. Therefore, the deciding factor is the method of navigation. Mobile platform is to navigate on a planar surface in an uncoupled fashion which means it should have three degrees-of-mobility as stated in the design criteria. The conceptual design that use the differential drive system and the Ackerman steering system are eliminated due to this factor. Conceptual design utilizing omni-directional wheels still remain as possible candidates since they both have three degrees-of-mobility.

Fault tolerance can be achieved for the three omni-directional wheel drive by having reserve actuators for each of the wheel actuators. This is a solution in joint level fault tolerance. The shortcoming of this concept is that the design requires a total of six servos.

The concept with four omni-directional wheels has fault tolerance in link level by design. Hence, fault tolerance is achieved by only using four servomotors instead of six. As concepts are evaluated for mobility needs and fault tolerance features, four-omni-directional wheel drive system is selected as the final design as shown in Figure B.17.

After selecting the drive system, sensors are required to be selected to obtain the information about the work environment of the platform. This platform is designed to be constructed for force-reflecting teleoperation experiments, which require force feedback signals from the slave system. The force feedback information is proposed to be created with respect to the proximity of the platform to the obstacles. Range sensors are to be used to measure the proximity to the obstacles on all four sides of the platform.

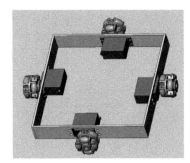

Figure B.17. Final version of the drive system design

A variety of range sensors are available commercially such as ultrasonic range sensors, infrared range sensors and laser range sensor. Fault tolerance can also be employed for the sensor configuration. Triple Modular Redundancy [42] configuration is to be used for sensors. Hence, three sensors can be placed on each side of the platform to form a triple voting configuration for the same sensory information. This means that a total of 12 sensors (3 sensors on each of the 4 sides) are to be used on the platform.

The control board for the platform should control at the four continuous rotation R/C servomotors and should have at least twelve input ports to receive the sensory information. It should also be able to communicate with the master system via a cable-free communication. Bluetooth communication protocol is selected for the wireless communication. Hence, the control board should also have a USB or RS-232 port for the Bluetooth connection.

Servomotors selected should be velocity controlled since the master system of teleoperation sends velocity signals to control the mobile platform.

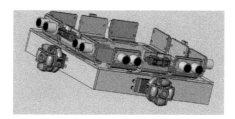

Figure B.18. Final version of the drive system design with sensor configuration

B.5. Part Selection

Parts to be used for the mobile platform construction are briefly described in the previous section. These parts are omni-directional wheels, range sensors, continuous rotation R/C servomotors, control board, side and top brackets, and Bluetooth device for wireless communication. The selection process of these parts is given below.

- Omni-directional Wheel Selection

There are two general types of commercially available omni-directional wheels. One is the Kornlylak Corp.'s "Transwheel®" as shown in Figure B.3. There are two producers for the second type of omni-directional wheels. Kornylak Corp.'s "Omniwheel" and NARP's "All-side Roller" has similar designs. "Omniwheel" and "All-side Roller" have a more sophisticated design to ensure the smoothness of the motion. Considering this, either "Omniwheel" or "All-side Roller" may be selected as the omni-directional wheel.

- Range Sensor Selection

Possible range sensors for this size of mobile platform and its application should have relatively good accuracy within one foot and should be reasonably priced with respect to the other parts of the system. There are two types of sensors available that meet these requirements; one is the Infrared Range Sensor (IRS) and the other one is the Ultrasonic Range Sensor (URS) as shown in Figure B.19. Different brands and types of these are available in the market in various prices.

Combination of these sensors can be used for the construction of the mobile platform. A total of twelve range sensors are to be used for fault-tolerant design. Eight of them can be selected as one type and the remaining four can be selected as the other type of range sensors.

Figure B.19. Ultrasonic range sensor and infrared range sensor from Parallax [133]

- Servomotor Selection

Continuous rotation R/C servos are available in various specifications of torque and energy consumption. Futaba® and Hitec are the most well-known producers of these servomotors (Figure B.20). These servomotors are mostly velocity controlled which meet the specifications of the teleoperation.

Figure B.20. Hitec [134] and Futaba [135] continuous rotation servomotors

229

- Control Board Selection

Possible control boards that are commercially available are reduced to three after examining their specifications against the design requirements denoted in the previous section. The first board is Pontech's SV203 board with 8 servomotor outputs and 5 inputs with 8-bit A/D converters as shown in Figure B.21. It has a serial port output for PC connection, which can also be connected to a Bluetooth device. Although the board meets the connection and servomotor port number specifications and it is fairly priced ($60), it does not have sufficient analog inputs for the twelve sensors. Three of these should be used at the same time to meet the specifications, which increase the board price to $180.

Figure B.21. Pontech's SV203 control board [136]

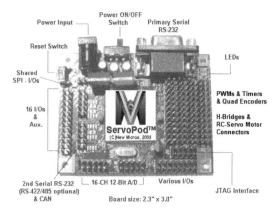

Figure B.22. NewMicros' ServoPod™ control board [137]

The second board is the ServoPod™ from NewMicros (Figure B.22). NewMicros has two versions of the board as serial port ($200) and USB ($250) connection boards. Both versions have 26 servomotor ports and 16 inputs with 12-bit A/D converters, which is more than enough for mobile platform specifications.

The last board is the Board of Education with Basic Stamp 2 from Parallax as shown in Figure B.23. This card has also a serial port for PC connection. It also has sixteen input/output ports that can be either used to drive servomotors or receive sensory information. This is sufficient to operate four servomotors and receive information from twelve range sensors as the initial design of the platform calls for. Board of Education with Basic Stamp 2 is also reasonably priced at $120.

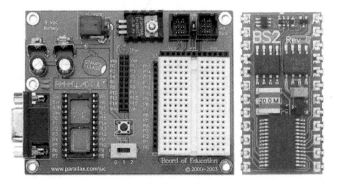

Figure B.23. Board of Education and Basic Stamp 2 from Parallax [133]

- Side and Top Brackets

Brackets are designed using SolidWorks. 3 mm aluminium sheets are used to manufacture the side brackets. Plexiglass material is used to manufacture the top bracket to better visualise the cable connections.

- Serial Port Bluetooth Device Selection

Although there are many producers of USB connection Bluetooth devices, RS-232 serial port connection Bluetooth device producers are limited. Gridconnect's Firefly was found to be most cost-effective. It can transfer data from 9600 to 115200 Baud Rates and its range is 330 feet. When two Firefly devices with the same baud rate are used at both ends (PC side and Control Board side), they are automatically connected without any need for software or driver instalment. The Firefly Bluetooth couple is shown in Figure B.24.

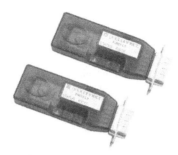

Figure B.24. Gridconnect's Firefly couple [138]

- Cost Estimate

Configuration of the sensors is selected to be eight of URS and four of IRS. The most cost-effective selection is made for the other parts that have two or more possible selections. The component costs are summarized in Table B.1.

Table B.1. Cost estimate of the mobile platform

Item	Unit Price	Quantity	Cost
Wheels	$12	4	$48
URS	$25	8	$200
IRS	$12	4	$48
Servomotor	$7	4	$28
Basic Stamp 2	$120	1	$120
Firefly	$80	2	$160
Misc.	$20	1	$20
Total			$624

B.6. Motion Planning

Servomotors have closed-loop velocity control of their own. PPRK of CMU is operated with three standard servomotors. Therefore, it can be assumed that these servomotors have enough power to operate the vehicle and calculation of the actuator dynamics is not necessary. No slip model is considered for simplification purposes since the platform is small and the motion will be at respectively low speeds.

Initially, linear motions of the mobile platform are required to be used in teleoperation experiments using the two-DOF gimbal-based master joystick. The orientation of the mobile platform is to be kept constant for this type of telemanipulation. On the other hand, when the system is controlled with the other master systems that have more DOF than two, all the capabilities of the mobile platform can be utilized.

- Regular Motion Planning

The motion of the vehicle is designated to be along x and y directions and the orientation must not change during these motions for teleoperation with gimbal-based joystick. When all four wheels are actuated, the orientation of the vehicle should not change, $\omega_v = 0$. Figure B.25 shows traction force T_i components and link lengths L_i of the platform.

The platform is designed so that every wheel is placed at the same distance from the rotation center of the vehicle.

$$L_1 = L_2 = L_3 = L_4 \qquad\qquad (B.1)$$

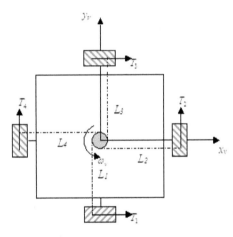

Figure B.25. Traction forces of the mobile platform with four-wheel configuration

Therefore, the same traction force should be applied to parallel wheels during the motion not to change the orientation of the vehicle.

$$T_1 = T_3 \ ; \ T_2 = T_4 \qquad\qquad (B.2)$$

If the applied traction force of one parallel wheel set is not equal to each other, compensation with the other wheel set can be achieved using the equation below and the orientation can still be kept constant.

$$T_1 - T_3 = T_4 - T_2 \qquad (B.3)$$

If the orientation of the platform is required to be changed without any linear motion, then the traction forces can be regulated as shown below.

$$T_1 = -T_3 = -T_4 = T_2 \qquad (B.4)$$

Using a master system with at least three DOF, linear and angular motion demands can be sent to the platform. The platform has to respond to these demands by having linear motion while changing its orientation. This can be achieved by applying unequal traction forces along the parallel wheels.

- Three-Wheel Motion Planning

When one of the four wheels fails, the mobile platform is required to continue its task as demanded by the master system. As a matter of fact, the only three omni-directional wheels are necessary to accomplish any planar motion demand. Nevertheless, the regular motion planning should be modified.

The failing wheel now acts as a pivot point as the other wheels are actuated as shown in Figure B.26. Since the failing wheel acts as a pivot point, the angular rotation about the pivot (ω_P) should be kept zero in order to keep the orientation constant. Therefore, when the parallel wheel of the failing one is active, traction forces should be set as:

$$2T_4 = T_1 - T_3 \qquad (B.5)$$

As the traction forces are regulated as shown in (B.5), motions along the x and y-axis can be achieved without any change in the orientation of the platform. The orientation of the platform can also be changed while the platform is moving along any axis by selecting the traction forces so that the equality given in (B.5) becomes invalid.

234

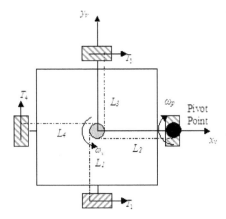

Figure B.26. Traction forces of the mobile platform with three-wheel configuration

INTERFACE SOURCE CODES FOR THE HOLONOMIC MOBILE PLATFORM

C.1. Source Code for Basic Stamp 2 (PBasic)

'This source code receives servomotor demands from the master system through Matlab and sends back
range sensor readings to the master system. Serial port is used for communication between the BS2 and PC.
' {$STAMP BS2}
' {$PBASIC 2.5}

'Sonar sensor ports are assigned in the portion below

```
PingXF      PIN   8
PingXB      PIN   6
PingYR      PIN   5
PingYL      PIN   7
Trigger   CON   5              ' trigger pulse = 10 uS
Scale    CON    $200          'used for scaling sonar readings
RawToCm      CON   2257           ' 1 / 29.034 (with **)
```

'Variables used in the code are described below

```
IsHigh      CON    1
IsLow       CON    0
rawDist1       VAR    Word
rawDist2       VAR    Word
rawDist3       VAR    Word
rawDist4       VAR    Word
cm1         VAR    Word
cm2         VAR    Word
cm3         VAR    Word
cm4         VAR    Word
Inst VAR Word
motorf1 VAR Word
motorf2 VAR Word
motors1 VAR Word
motors2 VAR Word
```

'Servomotor ports are assigned below
```
servoS1 CON 13
servoF1 CON 12
servoS2 CON 14
servoF2 CON 15
```

'Serial port communication rate and port are assigned below
```
baudmode CON 84
serial CON 16
```

'Main program running continuous in a do-loop
'First sonar sensors are read
'These sensor readings are formatted to be send to master system through Matlab
'Servomotor demands are received from master through Matlab
'Servomotor demands are reformatted to drive the servos through the BS2
'Servomotors are driven with the reformatted motion commands

```
DO

GOSUB Get_Sonar                 ' gets sensor values
```

'Sensor readings are formatted to be sent to Matlab
```
cm1 = rawDist1 ** RawToCm
IF cm1 >19 THEN cm1=19
cm2 = rawDist2 ** RawToCm
IF cm2 >19 THEN cm2=19
cm1=cm1-cm2+20
IF (cm1=20) THEN cm2=50
IF (cm1<20 AND cm1>15) THEN cm2=40
IF (cm1<16 AND cm1>10) THEN cm2=30
IF (cm1<11 AND cm1>5) THEN cm2=20
IF (cm1<6) THEN cm2=10
IF (cm1<26 AND cm1>20) THEN cm2=60
IF (cm1<31 AND cm1>25) THEN cm2=70
IF (cm1<36 AND cm1>30) THEN cm2=80
IF (cm1>35) THEN cm2=90

cm3 = rawDist3 ** RawToCm
IF cm3 >19 THEN cm3=19
cm4 = rawDist4 ** RawToCm
IF cm4 >19 THEN cm4=19
cm3=cm3-cm4+20
IF (cm3=20) THEN cm4=5
IF (cm3<20 AND cm3>15) THEN cm4=4
IF (cm3<16 AND cm3>10) THEN cm4=3
IF (cm3<11 AND cm3>5) THEN cm4=2
IF (cm3<6) THEN cm4=1
IF (cm3<26 AND cm3>20) THEN cm4=6
IF (cm3<31 AND cm3>25) THEN cm4=7
IF (cm3<36 AND cm3>30) THEN cm4=8
IF (cm3>35) THEN cm4=9
cm1=cm2+cm4
```

'Receives servomotor commands from Matlab
```
SERIN serial,baudmode,[DEC Inst]
```
'Sends sensor readings to Matlab
```
SEROUT serial,baudmode,[DEC cm1]
```
'Calculates the motor commands to drive the servomotors
```
GOSUB ForthBack
```

'Servomotors are driven with the commands below
```
IF (motorf1=750) THEN
LOW servoF1
LOW servoF2
ELSE
PULSOUT servoF1,motorf1
PULSOUT servoF2,motorf2
ENDIF

IF (motors1=750) THEN
LOW servoS1
LOW servoS2
ELSE
PULSOUT servoS1,motors1
PULSOUT servoS2,motors2
ENDIF

LOOP
```
'End of Main Program

'Sonar sensor reading subroutines
```
Get_Sonar:
```

237

```
PingXF = IsLow                          ' make trigger 0-1-0
PULSOUT PingXF, Trigger                   ' activate sensor
PULSIN  PingXF, IsHigh, rawDist1            ' measure echo pulse
rawDist1 = rawDist1 */ Scale              ' convert to uS
rawDist1 = rawDist1 / 2                  ' remove return trip

PingYL = IsLow
PULSOUT PingYL, Trigger
PULSIN  PingYL, IsHigh, rawDist3
rawDist3 = rawDist3 */ Scale
rawDist3 = rawDist3 / 2

PingXB = IsLow
PULSOUT PingXB, Trigger
PULSIN  PingXB, IsHigh, rawDist2
rawDist2 = rawDist2 */ Scale
rawDist2 = rawDist2 / 2

PingYR = IsLow
PULSOUT PingYR, Trigger
PULSIN  PingYR, IsHigh, rawDist4
rawDist4 = rawDist4 */ Scale
rawDist4 = rawDist4 / 2

RETURN 'End of sonar sensor reading subroutine

'Servomotor command calculations for going back and forth
ForthBack:
IF  (Inst<40 AND Inst>30) THEN
motorf1=700
motorf2=800
Inst=Inst-30
GOSUB Sides
ELSEIF (Inst<50 AND Inst>40) THEN
motorf1=725
motorf2=775
Inst=Inst-40
GOSUB Sides
ELSEIF  (Inst<60 AND Inst>50) THEN
motorf1=750
motorf2=750
Inst=Inst-50
GOSUB Sides
ELSEIF  (Inst<70 AND Inst>60) THEN
motorf1=775
motorf2=725
Inst=Inst-60
GOSUB Sides
ELSEIF  (Inst<80 AND Inst>70) THEN
motorf1=800
motorf2=700
Inst=Inst-70
GOSUB Sides
ENDIF
RETURN

'Servomotor command calculations for going sides
Sides:
IF (Inst=3) THEN
motors1=700
```

238

```
motors2=800
ELSEIF (Inst=4) THEN
motors1=725
motors2=775
ELSEIF (Inst=5) THEN
motors1=750
motors2=750
ELSEIF (Inst=6) THEN
motors1=775
motors2=725
ELSEIF (Inst=7) THEN
motors1=800
motors2=700
ENDIF
RETURN
```

C.2. Source Code for Matlab (M-file)

```
%This source code first receives the servomotor demands from the master system through Matlab
%Then opens the serial port and sends these commands to BS2 to drive the mobile platform
%A while loop is formed to make sure the servo demands are received by the slave
%Inside this loop, sonar sensor readings are also received from BS2 through the serial port
%The sonar sensor readings are reformatted and then sent out to Matlab Simulink environment

%Input the following in Matlab Command Window prior to running the code
%global MP_Object
%MP_Object = serial('COM1','baudrate',9600);
%MP_Object.terminator = 'CR';
%fopen(MP_Object);

function y = mobileplatformactual(u)  %receives servomotor commands from the master
global MP_Object  %serial communication is established through the global object
D3 = 3;
D9=1;
pulse_w=round(u);  %servomotor demands are assigned to pulse_w variable

%Below is the loop to make sure BS2 receives servo demands
while (D3 < 5)
    fprintf(MP_Object, '%d\n',pulse_w);  %Sends servomotor demands to BS2
    D3=fscanf(MP_Object, '%d\n');  %Receives sensor readings from BS2
end

%Sonar readings are reformatted to be sent to Matlab Simulink environment
D5=(D3-pulse_w)/100;
if D5 > 0
    D9=D5;
end

y=D9;  %Sends sensor readings to Simulink
```

239

Wissenschaftlicher Buchverlag bietet

kostenfreie

Publikation

von

wissenschaftlichen Arbeiten

Diplomarbeiten, Magisterarbeiten, Master und Bachelor Theses
sowie Dissertationen, Habilitationen und wissenschaftliche Monographien

Sie verfügen über eine wissenschaftliche Abschlußarbeit zu aktuellen oder zeitlosen
Fragestellungen, die hohen inhaltlichen und formalen Ansprüchen genügt,
und haben **Interesse an einer honorarvergüteten Publikation**?

Dann senden Sie bitte erste Informationen über Ihre Arbeit per Email
an info@vdm-verlag.de. Unser Außenlektorat meldet sich umgehend bei Ihnen.

VDM Verlag Dr. Müller Aktiengesellschaft & Co. KG
Dudweiler Landstraße 125a
D - 66123 Saarbrücken

www.vdm-verlag.de

www.ingramcontent.com/pod-product-compliance
Lightning Source LLC
LaVergne TN
LVHW022305060326
832902LV00020B/3280